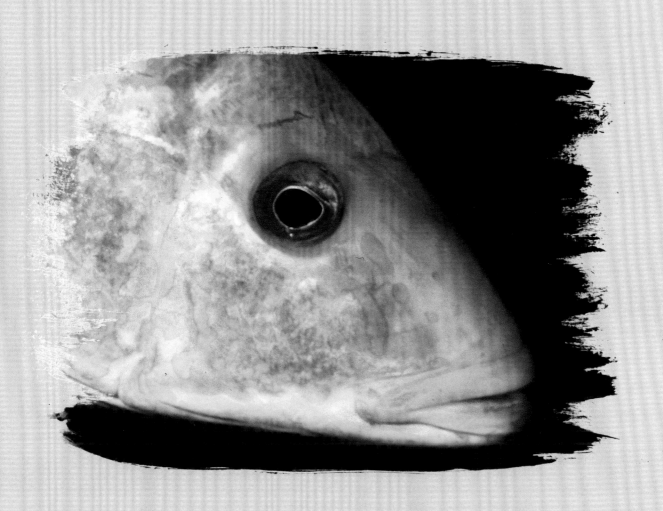

TANGANYIKA CICHLIDS
in their natural habitat

Ad Konings

Contents

Photo cover: A juvenile *Tropheus duboisi* at Karilani Island, Tanzania.

Photo back cover: *Tropheus* sp. "ikola" at Kekese, Tanzania

Photo endpapers: A school of mouthbrooding females and some males of *Cyprichromis leptosoma* at Isanga, Zambia.

Photo page 3: A male *Ophthalmotilapia ventralis* at Kambwimba, Tanzania.

Photos page 4: *Neolamprologus buescheri* at Isanga, Zambia, and *Julidochromis marlieri* at Halembe, Tanzania.

All photographs by the author unless otherwise credited.

ISBN 0-9668255-0-0

Cichlid Press
www.cichlidpress.com

Preface

The first edition of this book was published in 1988, and was the first of its kind to show many Tanganyika cichlids in excellent detail in their natural habitat, using photos taken mainly by Horst Dieckhoff. Over the ensuing decade it has helped many aquarists in designing their tanks and captive fish communities so as to enjoy these cichlids in all their splendour.

During those ten years copious observations, descriptions, and research on Tanganyika cichlids have yielded new insights into this complex group of fishes. Various scientists from Kyoto University in Japan have investigated the lives of many cichlids in great detail, and their findings continue to provide new surprises. Through anatomical studies of the lamprologines, Dr. Melanie Stiassny, of the American Museum of Natural History, has succeeded in creating a solid framework for the future classification and revision of this group, which comprises almost 40% of all Tanganyika cichlid species. Recently several scientists from Japan have discovered a number of information-bearing sections in the genetic material of the Cichlidae, with which they have been able to design, with confidence, a phylogenetic tree of many Tanganyika cichlids. And by making observations in the various cichlid habitats around the lake (except for the northwestern section between Kavala and Uvira), I myself have been able to gain a better insight into the nature of its complex fish communities, and gather a huge amount of information on the lives of these inspiring creatures. This book endeavours to integrate the findings of all these investigators into its descriptions of the various species and their habitats. In addition, in order to show the interested aquarist the splendour of Tanganyika cichlids, I have tried to photograph each species in its natural setting. Although each such photo is no more than a single frame in the long "film" of cichlid interactions, it says more than words can hope to, and gives the observer a direct view into the fish's life.

The observations in this book deal with the cichlids in their natural habitat, not in captivity. Although much of their natural behaviour is displayed in captivity, we must always remember that in the aquarium we are dealing with only a few individuals, and conclusions drawn from this small sample cannot safely be applied to the species in general. Nevertheless natural behaviour, as well as good health, is more likely to be exhibited if their habits and requirements are taken into account, and knowledge of these subjects is mandatory when creating a Tanganyika cichlid aquarium. It should be the the aquarist's goal to provide his fishes with the best artificial habitat, and only thus will he obtain the maximum enjoyment from them.

A number of questions that arose during the preparation of this book were quickly and efficiently answered by various ichthyologists, and I would like to express my gratitude to Drs. Melanie Stiassny, Jos Snoeks, and Tetsumi Takahasi, for sharing their knowledge with me. I would also like to thank Martin Geerts for sharing his encyclopedic knowledge of cichlids and his critical views on scientific literature. I am grateful to Mary Bailey for her meticulous correction of the grammar and content of the manuscript. Without the support and enthusiasm of Toby Veall (Rift Valley Tropicals, Lusaka, Zambia) and Laif DeMason (Old World Exotic Tropicals, Miami, Florida) various expeditions could not have been realized. Most of all I would like to thank my wife Gertrud, who supported my craving for in situ observations of cichlids, and was prepared to be left behind at home for several months, caring for family and fishes.

El Paso, November 1998

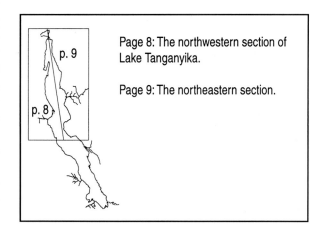

Page 8: The northwestern section of Lake Tanganyika.

Page 9: The northeastern section.

Kavimvira
Uvira
Luhanga
Makobola
Pemba (Bemba)

CONGO

Mboko

Lueba
Cap Banza
Cap Muzimu
Kiriza

Baie de Burton
Ubwari

Cap Caramba

Kisoshi

CONGO

Kabimba
M'toa
Kavala — Milima
Bendela
Kibige
Kalemie
Lukuga

Katibili

Rutuku

Kikonde

Cap Bendela

Cap Tembwe

Kitumba

Kambwebwe

Rusizi

Cap Muzimu (Ubwari)

Kisoshi

Lukuga (Kalemie)

Bujumbura

Ruziba

Magara

BURUNDI

Resha

Rumonge

Nyanza-Lac

TANZANIA

Kigoma

Ujiji

Malagarasi

Maswa

Cape Kabogo

Halembe

Magambo

Karilani
Bulu Point
Miyako
Kasoje
Luagala Point
Lubugwe Bay
Siyeswe Bay
Lumbye Bay
Mabilibili
Lyamembe
Sibwesa

MAHALE

Magara

Maswa

Kigoma

Halembe

Bulu Point

Karilani Island

Kasoje

Siyeswe

The lake and its cichlids

Lake Tanganyika lies in a roughly north-south position in the western arm of Africa's Great Rift Valley. Its coastline traverses four different countries with the Congo and Tanzania possessing the largest parts. The northeastern section belongs to Burundi and the southernmost shores lie in Zambian territory. The accessibility of the lake is restricted to a few localities where roads and lodging facilities have been built. These are found mainly in Burundi and Zambia.

Lake Tanganyika is the second deepest lake in the world (1470 metres) and its origin is thought to date back 9 to 12 million years. It in fact consists of three major basins, named the Zongwe (or Mpulungu), Kalemie, and Kigoma basins (Capart, 1949); according to Cohen *et al.* (1993), each of these three basins arose separately, with the central basin apparently the first to appear, 9-12 million years ago. The northern basin was formed 7-8 million years ago, while the youngest, the southern, is 2-4 million years old.

The shores of the lake are variable in their topography, and the different types alternate along the approximately 2000 km long coastline. The 3 basins are bordered by long stretches of rocky shore, i.e. (respectively) by the Marungu in the southwest, the Mahali Mountains on the central east coast, and the steep rocky coast on the western side opposite Kigoma. The shorelines between the three basins include rocky areas but consist in large part of sandy beaches. In the southern half of the lake beaches consisting of pebbles and small stones are common, especially where there are precipitous rocky coasts nearby.

River mouths are usually marked by a conspicuous growth of vegetation at the lake's edge. Most rivers and rivulets are of a temporary nature, carrying water only during the rainy season, but some larger, permanent, rivers have had (and still have) an impact on the range of rock-dwelling cichlids, frequently forming the boundary between adjacent but morphologically different populations. Important rivers are the Malagarasi (Tanzania), the Ifume (Tanzania), the Rusizi (Burundi), the Lukuga (Congo), the Lunangwa (Congo), and the Lufubu (Zambia).

The Lukuga River, on the west central coast, is the only outflow from the lake, at least at the present level of the water. In the past the water level has been periodically lower than currently, and at times the Lukuga may have flowed in the opposite direction, carrying water into the lake, or even have been dry. The Rusizi River, carrying mineral-rich water from Lake Kivu, contributes considerably to the present water level, and even more so to the lake's chemistry (Haberyan & Hecky, 1987). It has been estimated that the lake level rose 75 metres after the Rusizi was formed — approximately 10,000 years ago — between Lake Kivu and Lake Tanganyika (Coulter, 1991).

The precipitation during the rainy season (from October to December and from the end of February to April in the north, and from November to March in the south) raises the water level visibly and carries lot of sediment into the lake. Particularly in the northern end of the lake, the long rainy period causes frequent plankton blooms and hence low visibility in the normally crystal clear waters of the lake, which normally ranks as one of the clearest freshwaters in the world: visibility can be more than 20 metres. In sandy and muddy areas visibility is, of course, much reduced due to the suspended particles of sediment. Only on windless days is visibility here likely to exceed 10 metres.

Although the lake is very deep, oxygen is present only in the upper water layer. In the south the oxygen-bearing layer has a depth of about 200 metres, whereas in the north this layer is no deeper than approximately 100 metres. Mixing of the two layers is virtually non-existent apart from rare upwellings of the underlying anoxic layer in the south. Such upwellings, initiated by the prevailing southeasterly winds and subsequent cooling of the upper layer, may cause large numbers of fish deaths in the upper layer (Fryer & Iles, 1972).

TANZANIA

M A H A L E

Miyako
Kasoje
Luagala Point
Lubugwe Bay
Siyeswe Bay
Lumbye Bay
Mabilibili
Lyamembe
Sibwesa
Isonga

Kekese

Ikola
Karema
Ifume
Kilewa
Kalila

Kabwe
Cape Mpimbwe (Msalaba)
Utinta
Kapemba
Cape Korongwe

Nkondwe
Mvuna   *Kerenge*
Kipili
Ulwile   Mkinga
Kisambala
Mtosi
Namansi

Hinde B

Fulwe   Wampembe

TANZANIA

Kala
Kala
Mamalesa (Malasa)

Katili

Samazi
Muzi
Kasanga

Kambwimba
Kantalamba
Kalambo
Hore Bay
Isanga
Crocodile (Mutondwe)   Chisanza

Mbete (Kumbula)   Chituta
Mpulungu
Katoto   Kasakalawe   ZAMBIA

Sibwesa

Kekese

Mtosi

Msalaba

Mbete Island

Nkondwe Island

Wampembe

Kasakalawe

The fact that cichlids are so well represented in the lake may have a further cause in addition to their brooding behaviour: their feeding behaviour. The basic generalised cichlid is a bottom-oriented fish, which finds food and shelter in, on, and close to the substrate. The cichlid feeding apparatus is capable of evolving to suit almost any available food source, and this feature is thought to have played a major role in their evolutionary success in some areas. As well as the maxillary jaws, used to secure food, cichlids have a second set of jaws just in front of the gullet, the so-called pharyngeal apparatus, which consists of bony plates set with teeth suited to process the foods normally consumed. In piscivores these teeth are sharp and slender and designed to macerate the prey, whereas in molluscivores the teeth are round and stout and designed to withstand the pressure created by the pharyngeal bones on the shell of a snail during the crushing process. The maxillary jaws are likewise capable of modification to exploit different food sources.

Tanganyika cichlids have developed an astonishing array of feeding techniques, and this may be due to the long time the fish community has inhabited the lake. Cichlids living in the river systems of Africa show a much lesser degree of feeding specialisation; most of them are invertebrate feeders, planktivores, or piscivores. These types of feeding behaviour are seen in numerous species in the lake as well. Here, however, specialisations such as scale-eating, sponge-eating, algae-scraping, mud-eating, egg-eating, and super-specialisations of the fluviatile feeding techniques, have developed.

These specialisations among the cichlids have developed to such a degree that all species have a preference for a certain type of biotope and many of them are thus found restricted to such habitats. The evaluation of field observations and inventories of many local communities has resulted in the recognition of several types of biotope in which a specific cichlid community can be found. Each type of biotope has its own group of inhabitants which, on the whole, are not found in other biotopes; although some species may occur in different types of biotope these remain the exceptions.

A habitat, in the sense used in this book, is an environment in the lake which is large enough to provide space for a community of fishes. It includes the substrate as well as the water column above it or, in the case of a precipitous rocky coast, beside it. The splitting of a habitat into so-called micro-habitats is based on the very specific requirements of a few species. A niche is the environmental requirement of a single species. In the main part of this book I have arranged the chapters according to the different habitats recognized and tried to give a general picture of the communities found there.

If we combine cichlid feeding flexibility with their proven ability to colonise the entire coastline, down to the lower limit of the oxygenated water, then we can see that cichlids were the group best adapted to colonise the lake. The fact that most Tanganyika cichlid species are stenotopic (living in a restricted area) has greatly enhanced the speciation process among these fishes. Moreover, cichlids have been shown to be able to develop new species in a rather short period of time (Greenwood, 1965; Geerts, 1991), thus enhancing their ability to colonise any newly available habitat or niche.

Speciation

In recent years there have been important additions to our knowledge of the geological development of the lake itself (Scholz & Rosendahl, 1988; Tiercelin & Mondeguer (1991); Cohen *et al.*, 1993). It is of great interest that all investigations so far suggest that at certain times the lake level has dropped so much that three lakes (corresponding to the three basins) have existed instead of one. It has also been suggested that this has enhanced the speciation of the fishes (and other organisms) in the lake. Geographical isolation plays a key role in this process, and the existence of the three prehistoric sub-lakes is still reflected in the distribution of many cichlid species (e.g. Konings & Dieckhoff, 1992).

Cichlids are bottom-oriented fish and cannot normally cross deep open waters to settle in another area. Many species are dependent on the type of habitat for which they are "conditioned", and if such a type of habitat is interrupted by another, unsuitable, habitat, the latter may form a barrier in their distribution. The more widely suitable habitats are separated from each other, the less likely it becomes that such habitat-dependent species can migrate from one to the other. This is reflected in the distribution of many of the rock-dwelling cichlids in the lake.

Since cichlids need close contact with a substrate they can only disperse along the shoreline at depths of up to a maximum of 250 metres (deeper layers in Lake Tanganyika do not contain oxygen). Large parts of the Tanganyika shoreline are separated from the opposite side of the lake by very deep water which cannot be crossed by cichlids. Thus, if a cichlid on one side of the lake is to migrate to the opposite side, it needs to follow the contours of the shoreline. Hence the opposite side of a deep basin in the lake is the most distant point in terms of cichlid dispersal. Nevertheless, many cichlid species are found on both sides of the lake (but not at suitable habitats in between), and it thus appears they must have dispersed to the other side during a period of low water level. In the low level prehistoric lakes they were probably distributed along the entire shoreline, as they are in Lake Kivu (Jos Snoeks, pers. comm.). When the lake level rose they simply migrated upward, remaining all the time at their preferred depth (vertical migration). Even though the populations on the east and west sides of the lake are now isolated, they are still the same species; at least there are no apparent differences between them.

This is most beautifully illustrated by the distribution of *Tropheus annectens* (see figure page 68). It is noteworthy that *T. annectens* lives only near those locations on the present shoreline that were also inhabited during the period of low water level (steep rocky coasts). It is thus likely that *T. annectens* was present in such a prehistoric lake and did not disperse very far after the lake had risen to its current level.

The existence of more than a dozen new (at present not found in the prehistoric lake basins) species in the Zambian waters of the lake, which could not have been there at the time of the low water level because that entire area was dry land back then, suggests that species originate in new areas. It has therefore been suggested that most speciation in Lake Tanganyika takes place via the so-called Founder Effect (Konings, 1992; Konings & Dieckhoff, 1992).

When a new, suitable, habitat has been created (by the rising water level) it can be populated by any species able to cross the unsuitable habitat separating the "mother population" from the new area. Experiments show that most rock-dwelling species need several years before the first individual finds the new habitat (McKaye & Gray, 1984). Knowing that cichlids are able to produce offspring several times a year, it can easily be imagined that once a male and a female of a species have met at the new reef, they may produce several hundreds of offspring before the third individual of that species arrives.

Founder populations can thus be built up from a very few individuals and have a high probability of generating variation among their members as the result of inbreeding from a restricted gene pool. But, if inbreeding induces the creation of new species, then the large gene pool of the mother population will prevent the creation of new species, and instead actually stabilise the characters of the species. It thus follows that the majority of the variant populations and new species are to be expected in relatively new areas, and in the case of Lake Tanganyika these are the regions which were dry land when the water level was low.

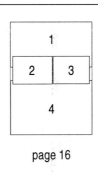

1. The author's camp at Halembe (Tanzania expedition 1997)
2. The village of Chisanza, Zambia, at sunset.
3. An adult chimpanzee grooming an infant at Gombe Stream National Park, Tanzania.
4. For an expedition on Lake Tanganyika one needs to take everything bar the kitchen sink (Tanzania 1995).

page 16

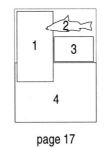

1. Kalambo Falls as seen from the Zambian side.
2. *Lates mariae* is a much-valued food fish.
3. Ndaga fishermen at Bulu Point, Tanzania.
4. Native fishermen are always helpful and proud to show off their fish (here *Hemibates stenosoma*).

page 17

Taxonomy and phylogeny

Lake Tanganyika cichlids were the subject of scientific study for the first time at the end of the 19th century. The most important taxonomist to work on them at that time was Boulenger, who described numerous new species from collections made by Moore, Lemaire, and Cunnington. Between 1909 and 1916, Boulenger published his four tomes of the *"Catalogue of the Freshwater fishes of Africa"*, in which more than 75 Tanganyika cichlid species were mentioned, almost all of which were described by him.

Apart from Boulenger and other well-known ichthyologists such as Pellegrin and Steindachner, each of whom described a handful of Tanganyika cichlids, a further major contribution to the taxonomy of these fishes was provided by Poll, who authored or co-authored the descriptions of more than 65 cichlid species and several revisions, and in 1956 published a very extensive expedition report on the cichlids of the lake, which is still considered the definitive publication on these fishes.

The natural environment

Until the advent of self-contained underwater breathing apparatus (SCUBA) most of our knowledge of Tanganyika cichlids was derived from morphological studies of preserved material and from local fishermen. In the sixties and seventies Brichard explored several parts of the lake with SCUBA gear and much of what we know of the natural behaviour of Tanganyika cichlids is derived from his observations (Brichard, 1978). He also collected many species new to science, most of which were described by Poll and later also by himself.

In the eighties it became obvious that the descriptions of yet more new cichlid species from Lake Tanganyika would make more sense if information about their biology and distribution could be supplied as well. And in fact the latest species descriptions, by Büscher (1989-1997), contain data on the lifestyle and natural environment of their subjects, and provide a more complete picture of the species in question.

Probably because of the availability of data other than those derived from morphological examination, purely taxonomic studies have lost some of their appeal, and the question has become more one of how all these cichlids are related, and, above all, of how such an assemblage could have originated in a single body of water. With the accumulation of information on the geology and chemistry of the lake more hypotheses about evolution have been put forward, and with the advancement of DNA technology some of these hypotheses have been put to the test.

Molecular taxonomy

While most DNA research (e.g. Sturmbauer & Meyer, 1993; Kocher *et al.*, 1995), has concentrated on determining the age of certain genes rather than investigating speciation processes, one type of DNA study has produced the most reliable phylogenetic tree of Tanganyika cichlids known to date. This is the work of Takahashi and his co-workers (1998) and involves the isolation and characterization of so-called SINEs (short interspersed repetitive elements) in the DNA of cichlids. Over the course of time such short elements are inserted at random sites in the fish's genome (DNA) by an extremely rare process, and once such an element is inserted it stays there forever, and is duplicated and carried over to the next generation. If such a species gives rise to a new species in the course of evolution, the new species too will have that element at exactly the same location in its DNA. So if two species have such a SINE inserted at exactly the same place in the DNA, it is reasonably certain that these two species have a common ancestor. If, however, the insertion site in a species does not have the relevant SINE, then this means that that species does not share an immediate common ancestor with the SINE-bearing species. Takahashi and his

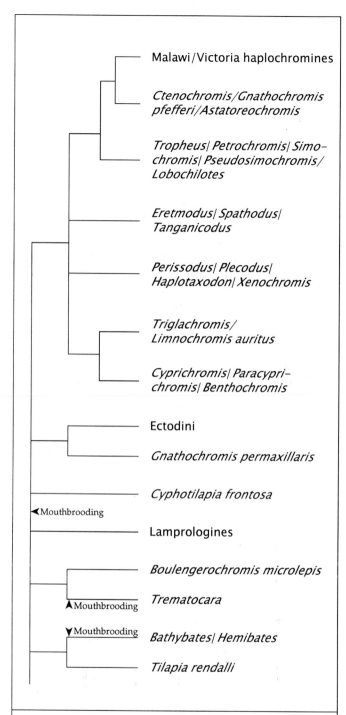

The phylogenetic tree of the cichlid genera of Lake Tanganyika derived from data published by Takahashi *et al.* (1998) and Nishida (1997).

The lamprologines include the genera *Variabilichromis*, *Lamprologus*, *Neolamprologus*, *Lepidiolamprologus*, *Altolamprologus*, *Julidochromis*, *Chalinochromis*, and *Telmatochromis*. The tribe Ectodini includes the genera *Asprotilapia*, *Aulonocranus*, *Callochromis*, *Cyathopharynx*, *Ectodus*, *Grammatotria*, *Microdontochromis*, *Ophthalmotilapia*, and *Xenotilapia*.

co-workers have isolated one such cichlid-specific SINE and found that all but one of the African cichlids investigated have copies of such an element in their DNA. Interestingly a species from Madagascar, *Paratilapia polleni*, does not have such a SINE, nor do *Etroplus maculatus* and two cichlid species from South America.

Although cichlid DNA can contain 2,000 to 20,000 of these elements, Takahashi and his co-workers (1998) have isolated and characterized just a small number of insertion sites, and 19 of these have proved informative in unravelling the phylogenetic tree of most Tanganyika genera (see diagram). Using this technique the authors have been able to estimate that mouthbrooding evolved at least four times in the Tanganyika species assemblage. The position and timing of such developments has been indicated in the diagram.

They have further found that all the species currently referred to as lamprologines are indeed members of a monophyletic (derived from the same ancestor) group, and have isolated 3 different loci (DNA insertion sites) which are unique to those lamprologines examined.

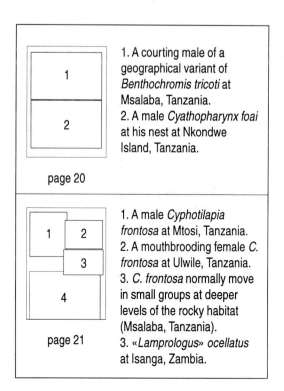

1. A courting male of a geographical variant of *Benthochromis tricoti* at Msalaba, Tanzania.
2. A male *Cyathopharynx foai* at his nest at Nkondwe Island, Tanzania.

page 20

1. A male *Cyphotilapia frontosa* at Mtosi, Tanzania.
2. A mouthbrooding female *C. frontosa* at Ulwile, Tanzania.
3. *C. frontosa* normally move in small groups at deeper levels of the rocky habitat (Msalaba, Tanzania).
3. «*Lamprologus*» *ocellatus* at Isanga, Zambia.

page 21

Unfortunately insertion of such repetitive elements in the DNA is such a rare event that up to now no informative loci could be isolated from the DNA of any Lake Malawi or Lake Victoria cichlid, so this technique cannot be applied to resolve the phylogenetic trees of those radiations. Interestingly the Victorian as well as the Malawian cichlids fall into the same group as *Ctenochromis* and *Astatoreochromis*, which suggests that the cichlids of both these lakes share a common ancestry with some of the Lake Tanganyika lineages.

Anatomy is not dead!

Although there is strong evidence that all lamprologines form a monophyletic group, until recently no further resolution of the intra-relationships of the different genera assigned to this group has been unequivocally determined. Recently, however, Stiassny (1997), in a state-of-the-art anatomical examination of almost all lamprologines known to date, elucidated many of the phylogenetic problems among lamprologines. Although her study was preliminary, Stiassny found a number of characters which are potentially informative in constructing a phylogenetic tree and which can be applied to classify these species.

She found that *Lamprologus moorii* has the most basal configuration of certain bones associated with the shoulder blade and those found around the eye, and therefore placed this species apart from all other lamprologines at the base of the phylogenetic tree.

Colombé & Allgayer (1985) had already assigned this species to a new genus, *Variabilichromis*, which was subsequently rejected by Poll (1986). Stiassny has now reinstated this monotypic genus. Colombé & Allgayer also assigned another species, *toae*, to a monotypic genus, *Paleolamprologus*, but although Stiassny found a derived character in this species there was nothing to support a separate position for it.

Stiassny (1997 and in prep.) further agrees with Sturmbauer *et al.* (1994) that the fluviatile lamprologines comprise a monophyletic assemblage within the group as a whole, and are derived from an ancestor that lived in Lake Tanganyika. The name *Lamprologus* — the type species is *L. congoensis* — is thus now restricted to the riverine species and the genus has no representatives in Lake Tanganyika.

In her as yet unpublished work Stiassny further confirms the monophyly of a modified *Lepidiolamprologus* and, in addition to *elongatus*, *attenuatus*, *profundicola*, and *kendalli* (*nkambae* is a synonym), she includes the following species in that genus: *meeli*, *hecqui*, *pleuromaculatus*, *boulengeri*, and *lemairii*. *L. cunningtoni*, which was hitherto assigned to *Lepidiolamprologus*, lacks some of the crucial characters of the group (see below) and is thus removed from this genus.

In addition to all the other informative characters Stiassny mentions in her paper, a single one was found that was sufficiently important to establish a new monophyletic group. This character involves the possession of an ossified (changed to bone) structure in the labial ligament. This ligament or tendon is found inside the lower jaw between the lefthand and the righthand sides, and in all other cichlids it contains a little knob of cartilage (like a patella but made of cartilage). In the species of the "ossified group" this knob is made of bone, a unique feature among cichlids. It appears that all species

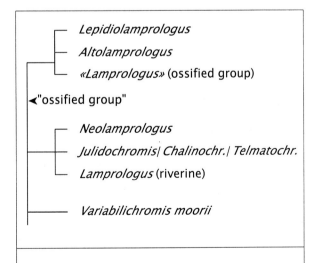

The phylogenetic tree of the lamprologines according to Stiassny (1997). The "ossified group" (see text for explanation) contains an assembly of species for which a generic name is not available. They are here temporarily classified as «*Lamprologus*».

now assigned to *Lepidiolamprologus* belong to the "ossified group" (as do numerous other species currently assigned to *Altolamprologus* and *Lamprologus sensu lato*) and now it is now clear that *L. cunningtoni* cannot be retained in that genus, because it lacks such an ossification.

Since *Neolamprologus tetracanthus*, the type species of the genus *Neolamprologus*, does not belong to the "ossified group", strictly speaking all lamprologines assigned to *Neolamprologus* should also lack such an ossification. However, as the intrarelationships of all these species have yet to be fully resolved, we are left with taxonomic and nomenclatural chaos. For the time being, apart from the species assigned to *Altolamprologus* and *Lepidiolamprologus*, all other members of the "ossified group" are assigned to «*Lamprologus*». The chevrons indicate that it is thereby understood that the species thus classified do not belong to *Lamprologus* in the strictest sense, but, due to the lack of another available name, are temporarily placed or retained in this genus. The species of the "ossified group" temporarily retained in or relocated to «*Lamprologus*» are: *brevis, callipterus, calliurus, caudopunctatus, finalimus, leloupi, multifasciatus, ocellatus, ornatipinnis, similis, speciosus, stappersi* (*meleagris* is a synonym, see page 207), and *wauthioni*.

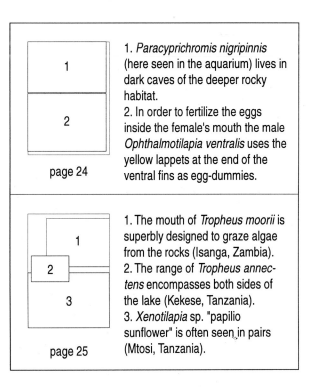

1. *Paracyprichromis nigripinnis* (here seen in the aquarium) lives in dark caves of the deeper rocky habitat.
2. In order to fertilize the eggs inside the female's mouth the male *Ophthalmotilapia ventralis* uses the yellow lappets at the end of the ventral fins as egg-dummies.

page 24

1. The mouth of *Tropheus moorii* is superbly designed to graze algae from the rocks (Isanga, Zambia).
2. The range of *Tropheus annectens* encompasses both sides of the lake (Kekese, Tanzania).
3. *Xenotilapia* sp. "papilio sunflower" is often seen in pairs (Mtosi, Tanzania).

page 25

The surge habitat

The upper three metres of the rocky and intermediate habitats harbour their own group of specialised cichlids. Several of these are herbivores which graze on the rich algal cover on the rocks. As far as its inhabitants are concerned, the main characteristics of the habitat are its relative abundance of food in combination with turbulent water.

A coastline with a steep inclination has a much smaller surge habitat than one which gradually shelves into deeper water. The surge habitat of shallow coasts can be more than ten metres wide, whereas a steep rocky coast, formed by huge boulders, has a very narrow surge area. Shallow shores can have a sandy or a rocky bottom; sometimes the sand is caked with minerals which have crystallized from the water, forming large soft-edged "slates" at the water's edge. The surge water is very well oxygenated owing to its turbulence, and at the same time the "faunal exhaust" (CO_2) is rapidly dispelled into the atmosphere, leading to a slightly higher pH in the surge zone than in other habitats. And by virtue of the continuous mixing of the top layer with the underlying water, the temperature remains rather constant, even in the extreme shallows.

Goby cichlids

The species of the three genera *Eretmodus*, *Spathodus*, and *Tanganicodus*, better known under their common name of goby cichlids, are adapted to the turbulent water of the surge habitat to such an extent that this is the only place where we find them. The goby cichlids are distinguished from other cichlids by their anatomy as well as their feeding and breeding behaviour. All members of the group have a similar anatomy: a short, laterally compressed, body with a remarkably long dorsal fin. To prevent them from being swept away by the surge their swim bladders are reduced and unable to keep them buoyant. When they rest on the substrate, i.e. while not swimming, their pectoral and pelvic fins are used to secure them in position between rocks, and it is common to see the hind parts of their bodies swaying in the current while their heads stay relatively still in the sometimes vigorously moving water.

The dorsal fin of goby cichlids is very long when compared to that of other Tanganyika cichlids. Cichlids use the soft-rayed posterior part of the fin to fine-tune their position in the water column, but in the case of the "Tanganyika Clowns", another common name for these cichlids, any movement of this part will push the fish closer to the substrate. This, of course, is exactly what they need when in turbulent water. The drawback is that when a goby cichlid wants to move from one place to another it can merely hop or energetically beat its pectoral fins. This odd-looking swimming behaviour is just one of the peculiarities of the goby cichlids that have made them popular among cichlid hobbyists.

Hopping and scurrying around in the habitat makes them rather vulnerable to predators, especially when we consider that they can't swim fast enough to dodge predatory attacks. As very few aquatic predators are able to swim in the extreme shallows — they are prone to being smashed against the rocks — most of the predatory pressure on goby cichlids comes from above - fishing birds are rather common around the lake! To defend against such attacks the dorsal fin of goby cichlids consists of a very large number of spines. In fact they sport the largest number among cichlids, sometimes up to 25! And the spinier the prey the less likely it is to feature on the menu of fishing birds.

Apart from the spiny dorsal, these cichlids also have a camouflage coloration consisting of vertical bars. If you ever have snorkelled in extremely shallow water you will probably have noticed the ever-moving pattern of reflections and shadows of the surface pattern on the substrate. In such an environment fishes with a colour pattern consisting of alternating light and dark bars are much more difficult to find than those with a solid colour.

Besides the high number of dorsal spines and reduced swim bladder, a third anatomical peculiarity is found in goby cichlids: bulging eyes. Because they spend most of their time sitting on the substrate, the placement of the eyes high on the head is advantageous. A high eye position allows the fish to observe its environment. The eyes cannot see the area just in front of the mouth, so they have not been adapted for visually selecting feeding spots (Yamaoka, 1997).

The various goby cichlids have different feeding specializations, although all species feed from the layer of algae and micro-organisms covering the rubble and rocks of their environment. This layer, called *Aufwuchs* (from the German meaning "growth on something") or bio-cover, consists of filamentous algae that are anchored to the rocky substrate, and which form a network onto which other algae, mainly unicellular algae or diatoms, attach. In many places in this algal mat small invertebrates, such as crustaceans, mites, and insects and their larvae, find refuge and food, and these organisms are preyed upon by some goby cichlids.

Breeding

All goby cichlids are mouthbrooders. There is almost no sexual dimorphism although the female is usually smaller than the male. Apart from *Spathodus marlieri*, goby cichlids are biparental mouthbrooders. The fact that male and female form a bond, sometimes for life, is quite unusual for mouthbrooders. The pair seem to defend a territory, a behaviour which is obvious in the aquarium, but sometimes difficult to establish in the wild because there is no way to be certain whether defended sites are indeed breeding territories or simply temporarily-defended feeding sites. Both male and female chase intruders from their premises, and most of their aggression is directed towards other members of their species.

In the aquarium spawning takes place on a horizontal surface, e.g. a flat-topped rock or the bottom of the tank. Egg-laying is preceded by the male moving over the spawning site while (probably) releasing his sperm. Assuming a slanting posture, leaning away from the male, the female moves slowly over the same site and quivers while she deposits one or a few eggs. She then turns around quickly to collect the egg(s). Next the male again quivers over the same spot while exuding his milt, but this time the female simultaneously "inhales" seminal fluid at his vent. The eggs are thus now fertilized inside the female's mouth, although some of them may already have been fertilized as soon as they were laid and touched the substrate.

About 20 to 30 eggs are produced per clutch. The female broods the eggs for the first ten to twelve days, but all the time the male stays close by her side. At the end of this period the female tries to get her partner's attention once more. At first it looks as if the pair is going to spawn again: both male and female are again active in chasing away all intruders. Once they have secured their "swopping site" the exchange of larvae can begin. The female starts shaking her head and releases one larva at the time to be carefully taken up by the male. The male is well aware of what is going on and waits impatiently with his mouth half open. Sometimes it looks like as if he is begging the female for another larva and almost picking them out of her mouth! After all the larvae have been transferred to the male's mouth they are brooded by him for another 7 to 10 days. Again the male and female stay close together during this period.

Once the male has released the fry they are ignored by both parents. For the first few days,

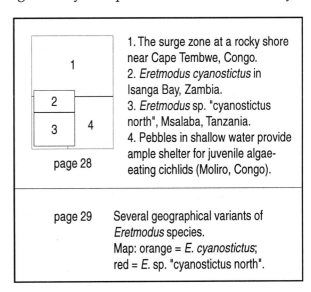

1. The surge zone at a rocky shore near Cape Tembwe, Congo.
2. *Eretmodus cyanostictus* in Isanga Bay, Zambia.
3. *Eretmodus* sp. "cyanostictus north", Msalaba, Tanzania.
4. Pebbles in shallow water provide ample shelter for juvenile algae-eating cichlids (Moliro, Congo).

page 28

page 29   Several geographical variants of *Eretmodus* species.
Map: orange = *E. cyanostictus*;
red = *E.* sp. "cyanostictus north".

① *E. sp.* "cyanostictus north", Magara

② *E. sp.* "cyanostictus north", Lumbye

③ *E. sp.* "cyanostictus north", Kapemba

④ *E. sp.* "cyanostictus north", Msalaba

⑤ *E. sp.* "cyanostictus north", Kapampa

⑥ *E. cyanostictus*, Ulwile Island

⑦ *E. cyanostictus*, Kasanga

⑧ *E. cyanostictus*, Chaitika

at least in *Eretmodus*, the young exhibit colour variation: roughly one half of the brood are dark brown while the other half have a light-brown coloration. On one occasion the two colour forms were grown on separately and it was found that the dark ones were males and the lighter ones females. It is not known whether this is true for all broods and if so what the reasons might be. Maybe there is no reason at all! However, a small school of dark- and light-coloured youngsters blend well into an environment consisting of a sun-lit bottom in shallow water.

According to Kuwamura *et al.* (1989) *Spathodus marlieri* is a maternal mouthbrooder which means that only the female broods the eggs and larvae. The male *S. marlieri* is considerably larger than the female — 10 cm vs 6 cm — and mouthbrooding females are found in very shallow water together with juvenile goby cichlids. *S. marlieri* roams through the habitat and does not seem to have a home territory. The other goby cichlids are monogamous and live in relatively small areas with a diameter of about two metres. The sequential biparental mouthbrooding of *E. cyanostictus*, *S. erythrodon*, and *T. irsacae* is sometimes considered very advanced by aquarists but in fact it probably represents a stage in between substrate brooding and maternal mouthbrooding. It is important that the fry, once released, are large enough to evade predators on their own. This means large eggs, and furthermore, that they need to be brooded for at least three weeks. The fact that the female is able to eat after a ten or twelve days' period of mouthbrooding is certainly an advantage, but the price is high: the totally defenceless larvae are exposed to the outside world during the transfer from female to male. Secondly the parents have to stay close together or else the female has to brood the larvae the entire period herself (which will happen if you isolate the mouthbrooding female). Most important is the fact that in principle any male can become a partner with any female; in other words there is no continuous male competition. Males of many maternal mouthbrooding species have to fight for a good position in the breeding arena and are selected by females for their fitness. A successful male may thus mate with several females. The latter system may assist in enabling a species to adapt more quickly to a changed environment.

Fortunately the habitat of Tanganyika cichlids has not changed much in the last few millions of years and seemingly less advanced breeding strategies are still very effective — advanced doesn't always mean better!

The goby cichlids are currently divided into three different genera which are characterized solely on the basis of tooth structure: spatula-shaped teeth are found in *Eretmodus* species, cylindrical teeth in *Spathodus* species (see photo page 32), and slender, pointed, teeth in *Tanganicodus irsacae*. Recent investigations using the DNA of all of these species (Verheyen *et al.*, 1996) have suggested that a generic separation of the goby cichlids on this basis may be artificial and that they would be better accommodated in a single genus — in this case *Eretmodus*, because it is the oldest of the three.

Eretmodus

Eretmodus species live on a diet of algae which are scraped from the rocks. They are usually found grazing from the upper sides of small rocks. The underslung mouth of the northern species, *E.* sp. "cyanostictus north", enables it to feed in a more horizontal position so that it can scrape algae from rocks in very shallow water without losing its balance. The teeth of *Eretmodus* closely resemble (rows of) tiny chisels; they are not pointed as in most other cichlids. These teeth are very effective at scraping the rocks clean of any algae attached to them, and scrape marks on a rock are a good indication that *Eretmodus* has been there. In fact these teeth are so peculiar that they are the basis on which *E. cyanostictus* was placed in a genus of its own. Such a taxonomic distinction seems outdated now, and it would not be surprising if, in a future revision, all goby cichlids were placed in the same genus, i.e. *Eretmodus*.

The teeth on the pharyngeal jaws, the bony plates just in front of the gullet, are numerous and close-set, but not as numerous as in other algae-feeding cichlids. Herbivorous cichlids use their pharyngeal teeth to break the algal cells so that the digestible cell content is released. Algae-eaters always have a long intestinal tract because plant material is difficult to digest and needs a long processing time in the gut. The intestine of *Eretmodus* is twice to three times as long as the fish itself, but other algae-feeders, such as *Petrochromis*, have guts with a length of up to 10 times their body length. A considerable quantity of sand grains has been found in the stomachs of wild-caught *Eretmodus*, and it is possible that the sand has a digestive function, as in seed-eating birds where ingested grit is used to grind the seeds in their gizzards. It is not known whether or not the sand is ingested deliberately or is merely picked up accidentally along with the algae. In the aquarium *Eretmodus* has not been seen deliberately picking up sand, but on the other hand it is sometimes difficult to keep *Eretmodus* in a tank without a sandy substrate.

Geographical variation

Eretmodus (type species *E. cyanostictus*) was described by Boulenger in 1898 from specimens caught near Mpulungu, which was then known as Kinyamkolo. In the southern part of the lake it is the only goby cichlid genus present, but shares the northern two-thirds of the lake with the other genera, represented by three species: *Spathodus erythrodon, Tanganicodus irsacae* and *Spathodus marlieri*. However, one rarely finds more than two different goby cichlids at the same locality. If there is more than one species at a site, one of them is an *Eretmodus* and the other may be either *T. irsacae* or a *Spathodus* species.

95% of the *Eretmodus* we keep in our tanks originate from Burundi; only a small number have been exported from Zambia. It is unlikely that the Burundi population is conspecific with *E. cyanostictus*. The fact that it shares the habitat with three other species may have "forced" it into developing a different feeding strategy or habitat preference. If we compare both forms carefully we notice that *E. cyanostictus* has a more terminally positioned mouth than the Burundi form. This difference seems to be con-

sistent in all populations and I consider these two forms to be two different species, and I have called the northern form *Eretmodus* sp. "cyanostictus north". It is in any case wise not to keep both forms/species in the same aquarium.

Several geographical morphs of *Eretmodus* can be distinguished in the lake. The difference between these variants is not as dramatic as that seen in some of the other mouthbrooding species of the lake, but it is nevertheless advisable likewise to keep these variants apart in the aquarium.

E. cyanostictus at Mpulungu and in the surrounding area are characterized by a number of bright blue dots on the head and upper half of the body, and this characteristic is found in all populations in the southern part of the lake between Kipili in Tanzania and Moliro in Congo. I believe that the taxon *E. cyanostictus* should be restricted to this group of populations. This is also the only species of this group that throughout its range does not share its preferred habitat with any other goby cichlid. The northern species, *E.* sp. "cyanostictus north", occurs in several different geographical variants. The most common one in the hobby originates from Nyanza Lac in Burundi and has 7-9 very

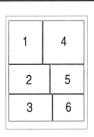

1. *Spathodus erythrodon* at Kapampa, Congo.
2. *Spathodus erythrodon* at Cape Tembwe, Congo.
3. *Spathodus marlieri* (from Burundi) in the aquarium.
4. Close-up of the teeth of *Spathodus erythrodon*.
5. *Spathodus erythrodon* at Kavala Island, Congo.
6. Head of *Spathodus marlieri* (from Burundi).

page 32

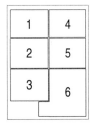

1. *Tanganicodus irsacae* at Kibige Island (Kavala, Congo).
2. *T. irsacae* at Kapampa, Congo.
3. *T. irsacae* in the aquarium (from Nyanzan Lac, Burundi).
4. *T. irsacae* at Halembe, Tanzania.
5. *T. irsacae* at Mabilibili, Tanzania.
6. The surge habitat at Nyanza Lac, Burundi.

page 33

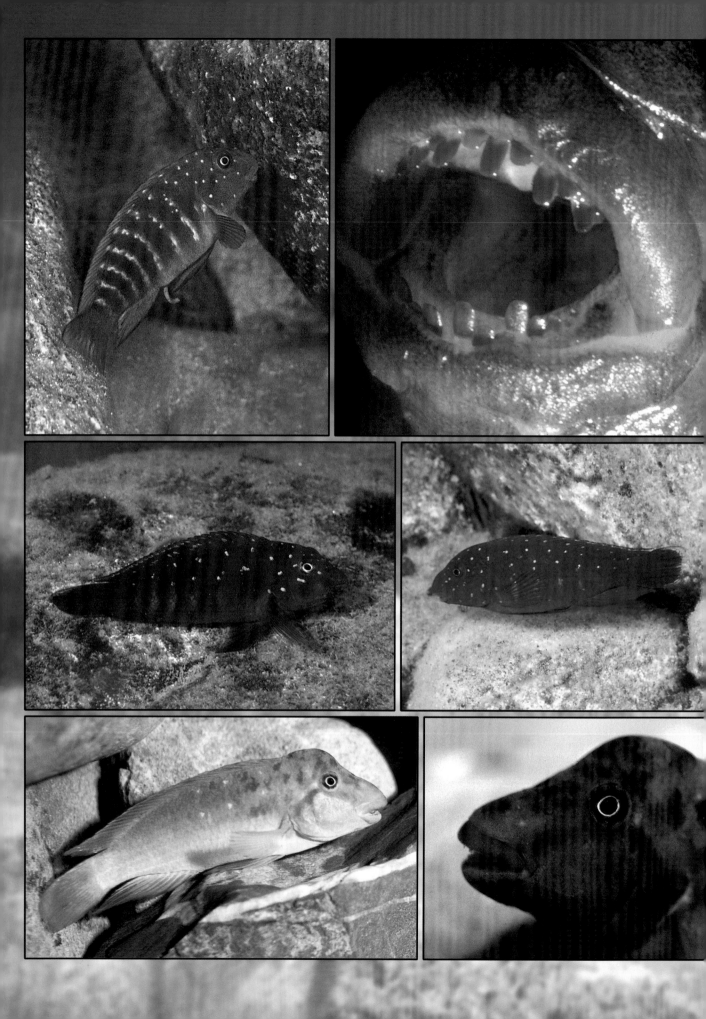

distinct vertical bars which are also visible on the upper part of the body and on the caudal peduncle. It has a few blue spots, mainly on the head. This variant is found in Burundi and along the northern shores of Tanzania, across the Malagarasi river mouth to Halembe. I have no pertinent data from populations in the north-western part of the lake but it seems likely that this variant is also found there.

The barring pattern of the local variant of *E. sp. "cyanostictus north"* at Kapampa in the Congo resembles the Burundi variant but in addition exhibits blue dots on the upper half of the body.

Along the Tanzanian coast we find several different forms/species of *Eretmodus*. The rocky coast north of Kipili, at Kapemba, is inhabited by a variant which shows hardly any barring at all but has blue dots on the head and upper body, although the dotted area does not extend to the caudal peduncle. I have not seen any other goby-like cichlid at Kapemba. Only about 4 km north of Kapemba begins the rocky area of Msalaba, the northern tip of which is better known as Cape Mpimbwe. Here lives a rather different form of *E. sp. "cyanostictus north"* which resembles the Burundi form in barring but lacks most of the blue colour on the chin. It does, however, have some pretty blue dots on the head which make this variant very attractive. Yet another variant occurs along the shores of the Kungwe mountain range. A photo of this variant appeared on page 17 of the first edition of this book, incorrectly captioned "*Eretmodus* from Karema". This variant is characterized mainly by a dark colour on the cheek and gill cover and by the lack of blue spots on the head.

Eretmodus species attain a total length of about 9 cm in the wild; the females remain about one fifth smaller. *Eretmodus* variants are seen mainly in pairs, and at some places many specimens can be observed at one time. *Eretmodus* are found at almost every rocky shore.

Tanganicodus

Tanganicodus irsacae is the smallest of the goby-like cichlids: males attain a total length of approximately 6.5 cm, females about 5.5 cm. Its diet consists of invertebrates picked from the substrate. The digestive tract is one and a half times to twice as long as the fish's total length. Examination of its contents has also revealed filamentous algae that may have been eaten along with the invertebrates that form the major portion of its diet (Poll, 1956).

At first glance *Tanganicodus* resembles *E. sp. "cyanostictus north"* and *Spathodus erythrodon*, but can be distinguished by its pointed snout and by a dark spot in the dorsal fin. A closer look reveals the long and pointed teeth which it uses to "winkle out" insect larvae and crustaceans from tiny holes in the substrate.

Tanganicodus has a wide distribution and is found at many places in the northern two-thirds of the lake. *T. irsacae* north of the Malagarasi river have a black spot in the dorsal fin, not in the soft-rayed part (Konings, 1988; Verheyen et al., 1995), but in the spiny part, approximately in the middle of the fin. I found dorsal spots in all populations of *T. irsacae*, but in the populations south of the Malagarasi these spots are not so distinct. The northernmost populations have a distinct spot sited more or less at the edge of the dorsal fin, whereas in specimens of other populations the spot blends with one or more of the vertical bars and is therefore located on the lower part of the fin. The members of the population at the type locality (Uvira, in the far north of the lake) have the spot at the edge of the dorsal, and the barring pattern is distinct on the lower part of the body but very faint or absent on the upper half. The population found at Kavala Island in the Congo has a dorsal fin spot on the edge of the fin but lacks the vertical barring. The southernmost *T. irsacae* population in the Congo is found near Kapampa. The individuals of that population have a distinct pattern of barring but have the dorsal spot continuous with one or two vertical bars. They are further characterized by vertically elongated blue streaks on the lower part of the body where there are just spots in other variants. The variants found at Halembe and along the Kungwe mountain range resemble each other. Both have distinct vertical barring on the lower half of the body and a dorsal spot continuous with one or two of the bars.

Spathodus

Spathodus erythrodon and *S. marlieri* belong to the third genus of goby-like cichlids. The teeth of *Spathodus* species are rather long and cylindrical (see photo page 32).

S. erythrodon is an algae- eater and only rarely found. Its digestive tract is twice to three times as long as the fish's total length (Poll, 1956). The males attain a maximum total length of about 7.5 cm. *S. erythrodon* looks very much like *E. cyanostictus*, and were it not for its dental morphology it would doubtless have been placed in that genus as well. Without investigating the shape of the teeth it is very difficult to identify it as being either *Eretmodus* or *Spathodus*. The northern populations of *S. erythrodon* lack the vertical bars of *Eretmodus* and hence it is fairly simple to distinguish the two species underwater, but along the Congolese shores *S. erythrodon* does have distinct bars on the lower half of the body (Poll, 1956). I have tried to identify them according to the position of the mouth: if the mouth is clearly underslung I have classified the "goby" as *Eretmodus* and if it is almost terminal I have called it *Spathodus* (this works only for those populations where we find both genera sympatrically, of course, because the true *E. cyanostictus* does not have an underslung mouth, only *E.* sp. "cyanostictus north").

S. erythrodon is exported mainly from Burundi and, infrequently, from Kigoma, but has a patchy distribution throughout the northern two-thirds of the lake.

S. marlieri is the largest of the goby cichlids and the least common, being found only in rocky habitats in the northern third of the lake. Males can attain a length of about 10 cm, and live in the lower part of the surge habitat. Unlike other goby cichlids, *S. marlieri* does not need close contact with the rocks and swims well away from the substrate (Brichard, 1978). Its diet consists of algae, invertebrates, and sand grains (inadvertently eaten?). A lot of diatoms and other unicellular organisms are undoubtedly ingested with the sand. In the aquarium *S. marlieri* has proved to be a rather nasty tankmate, tending to harass other species. All gobies are a little aggressive towards each other, but once they have formed a pair the battle is over. *S. marlieri* has been found to be a maternal mouthbrooder, and so there should be no need for the male and female to stay together.

Goby cichlids in the aquarium

Goby cichlids are best kept in pairs, or, in tanks larger than 600 litres, in groups of 6 to 8 individuals. As mentioned earlier they need very oxygen-rich water, a lush algae growth on the rocks, and the company of similar-sized cichlids such as *Tropheus* species. It is a good idea to have some fine sand on the bottom. Try to start with a group of juveniles and let them pair off. If only adults are available, select a large specimen and a smaller one which is full-bodied. Sometimes it is even possible to pick a "mated" pair from the shop tank. When two fishes of different size are "hanging out" together they are very likely a pair.

The origin of the goby cichlids

It is possible that the Tanganyika goby cichlids and some *Orthochromis* species, which are found in rivers surrounding the lake, have a common ancestor (Konings, 1988). *O. malagaraziensis* is found in the Malagarasi (a river

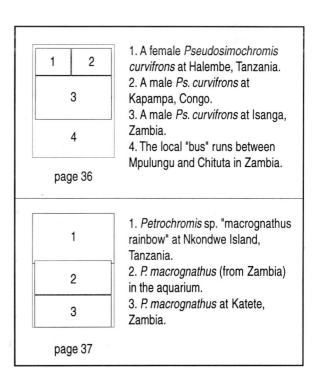

1. A female *Pseudosimochromis curvifrons* at Halembe, Tanzania.
2. A male *Ps. curvifrons* at Kapampa, Congo.
3. A male *Ps. curvifrons* at Isanga, Zambia.
4. The local "bus" runs between Mpulungu and Chituta in Zambia.

page 36

1. *Petrochromis* sp. "macrognathus rainbow" at Nkondwe Island, Tanzania.
2. *P. macrognathus* (from Zambia) in the aquarium.
3. *P. macrognathus* at Katete, Zambia.

page 37

east of the lake) and is a mouthbrooder living mainly in the torrents (Seegers, 1992). *O. polyacanthus*, found in river systems west of the lake, is a maternal mouthbrooder (Mary Bailey, pers. comm.), but a form which Seegers (1992) classifies as *O. spec. conf. polyacanthus* appears to be a facultative biparental mouthbrooder. Recently these *Orthochromis* species have been placed in *Schwetzochromis* (Kullander & Greenwood, 1997) but this re-classification has been found unacceptable by others familiar with these fishes (Bailey, pers. comm.).

In shape and in habitat preference these two *Orthochromis* species resemble Lake Tanganyika's goby cichlids. Male and female *Orthochromis* have the same colour pattern which at least suggests that they are not your regular maternal mouthbrooders and the male lacks ocellated egg-spots on the anal fin. None of the goby cichlids have such egg-spots.

Seegers also reports on *O. torrenticola* and shows a photo of male collected in the Congo. This species has egg-spots (although not ocellated) and overall bears more of a resemblance to the *Astatotilapia* species found in African rivers. This species, and probably also *O. machadoi*, may have derived from a different ancestor than *O. malagaraziensis* and *O. polyacanthus* and is not thought to share an ancestor with the goby cichlids from Lake Tanganyika.

DNA investigation (Takahashi *et al.* in Nishida, 1997; SINE insertion site analysis) shows that the goby cichlids are an older group than the *Astatotilapia/Ctenochromis/Tropheus* assemblage and it is thus possible that *O. malagaraziensis* and *O. polyacanthus* are derived from the lacustrine goby cichlids (instead of the other way around, i.e. the goby cichlids being derived from invading *Orthochromis* (Konings, 1988; Seegers, 1992)). But if either of these two species, in a future analysis, proves to belong to the *Astatotilapia* assemblage rather than to the *Eretmodus* group, then it would be impossible for these species to have given rise to the lake's goby cichlids.

Stiassny (1997) suggested that the riverine lamprologines are derived from the lacustrine and a similar scenario may have taken place regarding the goby cichlids and some *Orthochromis* species. But not until the SINE insertion sites (see page 18 for explanation) of these *Orthochromis* species have been determined can anything definite be said regarding the possible ancestral relationships of these two groups of taxa — and it may turn out that the goby cichlids are not related to these *Orthochromis* at all!

Pseudosimochromis curvifrons

Pseudosimochromis curvifrons resembles *Tropheus* species (see page 51) not only in morphology but also in feeding behaviour. Like *Tropheus moorii* it feeds exclusively on filamentous algae, but instead of effectively shearing away the algae, as in *Tropheus*, *P. curvifrons* grazes from the biocover by nibbling and cutting the strands (Yamaoka, 1983a; Takamura, 1984). Although this species is never found in large numbers, it is present at almost every rocky shore. Interestingly it has not developed distinct geographical races as in other rock-dwelling maternal mouthbrooders. In the northern part of the lake the males have a greenish hue on their mostly silvery body, whereas those in the south are a bluish colour. Females permanently exhibit a pattern of vertical barring which is also seen in male juveniles.

Males are territorial and their breeding territories can have a diameter of between 3 and 5 metres (Kuwamura, 1987a). Within their territories males are most aggressive towards other algae-feeders, in particular members of their own species, but also towards *Tropheus* species. Females are solitary and do not defend a feeding territory. Kuwamura (1987a) found that not all males occupied a territory and that non-territorial males may sneak into another male's territory and mate with "his" female during spawning (this behaviour is actually termed "sneaking"). Needless to say, such "sneak males" are vigorously chased from the premises, but Kuwamura found that on many occasions they were able to fertilize a few eggs. He also noticed that the pair would continue the spawning procedure (T-position technique) for 9 to 24 minutes after the last eggs were laid! This possibly suggests that eggs are still being fertilized after the female has collected them in her mouth. Moreover, non-territorial males were also observed sneaking after all eggs were laid; for them to have any effect the eggs should not have all already been fertilized in the female's mouth.

The maximum total length a male *P. curvifrons* can attain lies around 14 cm; females remain smaller at about 11 cm.

Petrochromis

Petrochromis is a successful genus whose members are the most important inhabitants of the rocky habitat. *Petrochromis* species are algal grazers which are superbly adapted for digesting vegetable material. Some species of this genus (e.g. *P. trewavasae*) have an intestine ten times the fish's length. Thus an adult *Petrochromis* may have a digestive tract measuring more than two metres in length! Not only the digestion of the food is well catered for, but also its acquisition. The broadly enlarged lips of *Petrochromis* support countless slender teeth. Most species feed by combing unicellular algae (diatoms) from the filamentous algae that are anchored to the rocks, simultaneously collecting anything which is not attached to the substrate, including sand grains (Yamaoka, 1997). Much of the biocover is collected by pressing the "toothy" lips against the substrate. The movable teeth adjust themselves to the contours of the rocks, and by closing its mouth the fish extracts loose material and diatoms from the algal lattice of the biocover. There is no other group of species in Lake Tanganyika whose members are able to collect algae in such an efficient manner.

One of the largest algae combers in the lake is *P. macrognathus* — a mature male can attain a maximum total length of about 21 cm. The mouth of this species is broad and protrudes downward. It lives in the upper two metres of the rocky habitat and mostly combs loose algae from the vertical faces of boulders in the turbulent water. In order to cope with the movement of the surge its feeding speed is fast. It is also a very shy species, probably because of heavy predation by birds and/or otters. Males are territorial and hide in large caves when threatened.

The holotype of this species was caught on the Congolese coast near Luhanga, 12 km south of Uvira, but it has a lake-wide distribution. More than a decade ago the so-called "Green Kabogo Petrochromis", discovered by Horst Dieckhoff near Cape Kabogo, appeared in the hobby. This form appears to be a race of *P. macrognathus*. *P. macrognathus* has since been found at other localities and has also been exported for the aquarium trade from the Zambian portion of the lake. *P. macrognathus* males from the north of the lake — the species has not been found north of Cape Kabogo along the eastern shores — have an overall dark-green colour or a pattern of vertical stripes, while the belly is usually more yellow. This variant also occurs in the central part of the lake.

South of the Mahali mountains, along the eastern side of the lake, a more colourful variant or species inhabits the upper two metres of the rocky habitat; for the time being I have named this form *P.* sp. "macrognathus rainbow". The mouth structure of the "Macrognathus Rainbow" closely resembles that of typical *P. macrognathus* and is unlike any other known *Petrochromis* species, which strongly suggests that both forms are conspecific. But since there are several "isolated" cichlid species along the central eastern shore of the lake, including endemic *Petrochromis* (see next chapter), I prefer to treat the colourful "Macrognathus Rainbow" as a separate species. In the southern portion of the lake a more yellowish variant is found, but this resembles the typical *P. macrognathus* and it is safe to regard it as a geographical variant of that species.

In 1986 Dieckhoff (pers. comm.) discovered the largest *Petrochromis* ever recorded: the as yet undescribed *P.* sp. "giant". He found it "at an

page 40	Some geographical variants of *Ophthalmotilapia ventralis* and their ranges: The "Black-&-White" Ventralis (black), the "Yellow Ventralis" (yellow), and the form at Moliro (orange).
page 41	The southern variants of *Ophthalmotilapia ventralis*, the so-called "Bright-Blue Ventralis" (blue).

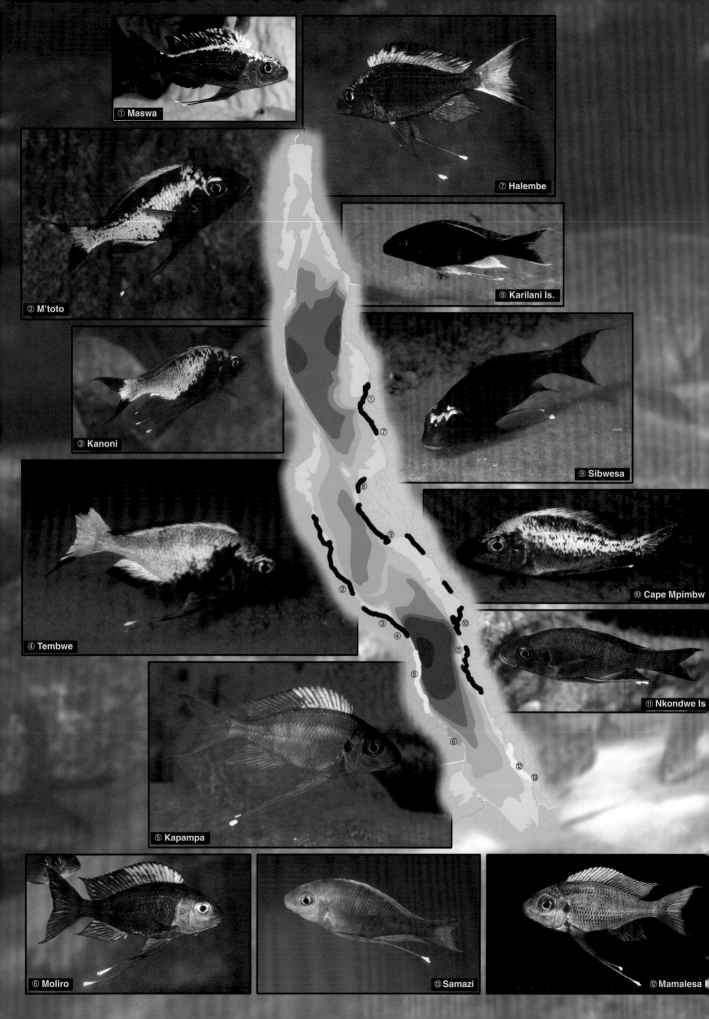

① Maswa

⑦ Halembe

② M'toto

⑧ Karilani Is.

③ Kanoni

⑨ Sibwesa

④ Tembwe

⑩ Cape Mpimbw

⑤ Kapampa

⑪ Nkondwe Is

⑥ Moliro

⑬ Samazi

⑫ Mamalesa

① Chimba

② Kachese

③ Cape Chaitika

④ Katoto

⑤ Chituta

⑧ Kambwimba

⑦ Kantalamba

⑥ Isanga

island near Kipili", but it was rediscovered in 1994 at Fulwe Rocks, a small rocky reef near Wampembe, about 80 km south of Kipili. The maximum size of this species is estimated at about 40 cm. Males are extremely territorial and disputes between them are commonly observed. *P.* sp. "giant" does not seem to be related to *P. macrognathus* as it has a terminal mouth with a slight overbite. It may therefore be more closely related to *P. polyodon* (see page 46). It occurs in the upper three metres of the water column among the huge boulders of the reef. Males have an overall green-blue colour; females seem to have a more yellowish colour with vertical bars (although none were collected for proper identification). Needless to say, all these large *Petrochromis* are very aggressive in their territorial defence and, unless accommodated with artificial rocky reefs of five metres diameter or more, they are not suitable for the home aquarium.

Featherfins

Featherfins is the collective common name for cichlids of the genera *Ophthalmotilapia*, *Cyathopharynx*, and *Cunningtonia*. Males of all these species are characterized by elongate pelvic fins, usually tipped with yellow lobe-like extensions (lappets). Usually only one species is found in the surge habitat: *O. ventralis*.

O. ventralis is regularly encountered in large populations and has a wide distribution encompassing the Tanzanian coast south of the Malagarasi River, the entire Zambian shore, and all along the shoreline of the Congo at least as far as Cape Tembwe, but may be found further north as well. There are many different geographical variants known, and many of these have been introduced into the hobby (see distribution maps pages 40 and 41). All the southern variants are known under the trade name "Bright Blue Ventralis", and the coloration of highly motivated males is indeed bright blue. Geographical variants can be distinguished by the variable black markings on the body. The "Bright Blue" variants of the south are bordered on both sides of the lake by the yellow forms found, on the eastern shore, around Mamalesa

Island in Tanzania, and, on the western coast, around Kapampa in the Congo. These yellow forms are in turn bordered to the north by populations with black-and-white-coloured males, again on both sides of the lake.

Male *O. ventralis* are territorial and most males defend a rather small patch or "corner" on a large rock in the upper metre of the rocky habitat. Their territories are rather large and have a diameter of between two and three metres. Within such territories only males of the same species are aggressively chased, and here too are females courted. Females, meanwhile, are gregarious and form schools of up to 500 individuals, which move inshore to feed on plankton and loose Aufwuchs.

The breeding technique of *O. ventralis* has been studied in captivity. The interesting part involves the function of the yellow lappets at the ends of the male's ventral fins: these resemble eggs, and function as egg-dummies in order to ensure fertilization of the real eggs inside the female's mouth. Males do not usually construct a conspicuous spawning site (nest), but sometimes a site is faintly marked by a thin patch of very fine sand on top of the rock. In captivity, however, *O. ventralis* will most likely construct sand-excavated spawning sites when no suitable rocks are provided. All intruders are chased from the immediate environment of the nest, and conspecifics are repelled with particular vigour.

The male courts females, when they visit his territory, by lead-swimming to the nest in a typical undulating fashion with the fins simultaneously folded against the body. The male continuously swims from the nest towards the females and "undulates" back. He reacts rather excitedly when a female responds to his advances, swimming up and down between the approaching female and the nest in a seemingly uncontrolled manner.

If the female proves to be interested, the male descends to the nest, and, while still quivering, releases some of his semen. He then stops quivering and lifts his tail from the nest, as if leaving; his pelvic fins, however, remain in contact with the substrate. The yellow lappets, which seem to be "swollen" during spawning, are main object of the female's attention. Since their shape and colour are exactly those of real eggs, she tries to pick them up. Sometimes the lappets disappear completely inside the female's mouth. At the same time, however, she collects the semen. At this point the male swims from the site and waits at a short distance. The female now deposits one or more eggs which she immediately picks up. She then leaves the spawning site, triggering the male into following her and trying to lead her back to the nest. The next batch of eggs may, however, be laid in another male's nest. And this time the female has some eggs in her mouth which will be fertilized during the next round.

Far from being purely decorative, the lappets at the ends of the male's pelvic fins are among the most realistic and effective egg-dummies found among cichlids. Long after all the eggs have been laid the female is still attracted by these dummies. This assures the proper fertilization of all the eggs.

When spawning is completed the female joins a school consisting mainly of brooding females. Aquarium observations suggest that the female eats small food particles during incubation. Probably in order to conserve energy and be as inconspicuous as possible, mouthbrooding females hardly move a fin. After about three weeks of incubation, the fry of all the brooding females in the school are released simultaneously in shallow water, where they immediately form a school. The fry stay close together and move around in the upper centimetres of the open inshore water. They often mix with fry of other species, even with juvenile killifishes (*Lamprichthys tanganicanus*). Since the fry schools are formed at intervals, the individuals of a single school are all about the same size, and the mingling of schools probably occurs only when the individuals of both groups are approximately the same size.

O. ventralis is a herbivore, and adults as well as fry feed from the plankton and from the biocover on rocks. The intestine has a length three to four times that of the fish, which confirms the vegetarian diet. The maximum total length of a male can be about 15 cm; females remain a few centimetres smaller.

In captivity *O. ventralis* needs plenty of space because males can be rather belligerent. In order to breed *Ophthalmotilapia* the tank should have a length of at least 150 cm. It is advisable to keep a juvenile male together with an adult male in a breeding group so that the dominant male displays his best colours almost permanently. The aquarium can be divided into a rocky part and a patch of open sand, or alternatively a single flat stone will suffice. These cichlids have no special requirements regarding food, although one should keep in mind that they are vegetarians, and a few pairs of lamprologines will make suitable tankmates.

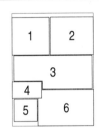

page 44

1. *Petrochromis* sp. "giant" at Fulwe Rocks, Tanzania; note the large round comb-marks left behind by this huge algae-comber.
2. *P.* sp. "giant" with two *P.* sp. "kipili brown" (Fulwe Rocks, Tanzania).
3. Two displaying males of *P.* sp. "giant" at Fulwe Rocks.
4. A small part of Fulwe Rocks near Wampembe breaks the surface.
5. Packets of snail eggs attached to a rock.
6. Camp at Wampembe, Tanzania.

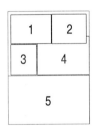

page 45

1. The crystallized minerals on the rocks provide abundant shelter for the smaller cichlids.
2. The freshwater jellyfish, *Limnocnida tanganyicae*, is sometimes found in great abundance.
3. The shady sides of large smooth rocks are commonly patrolled by *Neolamprologus furcifer*.
4. The crystallized minerals are continuously removed from the exposed sides of rocks but accumulate to form thick layers where browsing cichlids have no access.
5. A foraging school of *Petrochromis famula* at Mvuna Island, Tanzania. Only force of numbers enables them to feed in the territories of the more aggressive *Petrochromis* species; here *P. ephippium*.

The shallow precipitous rocky habitat

The rocks in this habitat are piled on top of each other, forming a complicated network of caves and crevices. The biocover on the rocks is virtually free of sediment, which allows the algae to flourish, in some places forming thick green mats. Because of this abundance of food, competition for living space is high, and only the strongest and most aggressive species are able to secure a territory.

Petrochromis

The most common group of species in this habitat is *Petrochromis*. *Petrochromis* species are the largest and most aggressive herbivores in the lake, and are found on every rocky shore. They are superbly suited to feeding on algae; some members of the genus (e.g. *P. trewavasae*) have an intestine ten times the fish's length. Thus an adult *Petrochromis* can have a digestive tract measuring more than two metres in length! Not only the digestion of the food, but also its collection, is optimally arranged. The broad, enlarged lips of *Petrochromis* support numerous slender teeth with three-pronged tips, and when the lips are pressed against the substrate the movable teeth fit into the contours of the rocks. Closing the mouth enables the tips of the teeth to comb loose material and also diatoms from the algal strands attached to the rock. The mouth is usually opened and closed several times at the same spot. *Petrochromis* species are very efficient at grazing the thick layer of biocover; important because, owing to their overall size, they need large quantities of food every day.

Petrochromis are maternal mouthbrooders, and males are strongly territorial. Their territorial aggression is directed not only towards conspecific males but against all *Petrochromis* species entering their feeding grounds. The feeding territory is smaller than the breeding territory but can still be more than two metres in diameter, whereas the breeding territory can have a diameter of more than five metres. Owing to the high competition for space in the upper rocky habitat, almost every inch is occupied by one or another *Petrochromis*. The territorial male must feed within his own territory, as if he left it to feed elsewhere, he would probably find it occupied by another male on his return.

On average *Petrochromis* females brood about 15 eggs per clutch. Spawning takes place in a cave or dark crevice in the male's territory, and the eggs are fertilized in the female's mouth as she nuzzles the male's anal fin and "inhales" the milt simultaneously released by him. The fry are released after about 30 days of incubation during which the female can take small food particles, either for herself or for her offspring in her mouth.

Males of several *Petrochromis* species have two different types of spots on the anal fin (and sometimes on the dorsal fin as well): yellow, ocellated, spots (egg spots) on the posterior part of the anal fin and non-ocellated spots on the anterior part. The non-ocellated spots are found on the tips of the membrane between the anal spines and may in fact function as egg dummies during spawning: the female tries to collect these fake eggs, and at the same time the eggs inside her mouth come into contact with the male's semen. The ocellated spots are also found on female anal fins, but here they are usually smaller and less numerous.

There are many species of *Petrochromis*, and at a single location it is possible to find up to 8 different species! At almost every rocky shore one can find a *P. polyodon*-like species, *P. famula*, a *P. orthognathus*-like species, a *P. macrognathus*-like species, *P. fasciolatus*, and *P. ephippium*.

Petrochromis polyodon

The most common species or species-group in this habitat is *P. polyodon* and its closely related forms. *P. polyodon* is a large herbivore with a maximum total length of about 20 cm. There are many geographical variants and forms (species?) that are morphologically very similar to *P. polyodon*, whose type material was collected near Mpulungu in the south of the lake, which means that the yellow-blue *Petrochromis* from the south is *P. polyodon* (for discussion see Herrmann, 1996). Along the southern shores of the Congo there is a very colourful variant/species resembling

P. polyodon, and for the time being this is treated as a geographical variant. *P. polyodon* is also found along the southern shores of Tanzania up to Kala. Along the central eastern shores there are several *polyodon*-like forms which have been given different names. Between Kipili and Karema we find the so-called *Petrochromis* sp. "texas kipili", and along the rocky shore north of Ikola there is a large dark blue *Petrochromis* which unfortunately I was unable to capture on film. Further north, along the Mahali mountain range, lives the elusive "Texas Petrochromis", which I found only at Mabilibili and Lumbye Bay. This species is so named because the blue-spotted pattern on the head resembles that of *Herichthys carpintis*, which is sometimes erroneously known as the "Texas Cichlid".

From Halembe to the northernmost part of the lake in Burundi we find the so-called *Petrochromis* sp. "kasumbe", which has been regarded as a race of *P. polyodon* (e.g. Kuwamura, 1986a), but which I believe to be a distinct species. Another species, from the Ubwari Peninsula in the Congo and different from the "Kasumbe Petrochromis", has been exported as *P.* sp. "texas red". Its name is derived from the fact that it looks like the "Texas Petrochromis" but has a red-coloured tail. It seems that the "Kasumbe Petrochromis" is the northern equivalent of *P. polyodon* in the south, and that there are several different forms in the central part of the lake.

Petrochromis orthognathus

A similar distribution pattern is also found in *P. orthognathus*-like species. In the northern half of the lake we find *P. orthognathus*, in the southern half *P.* sp. "orthognathus tricolor", and along the central eastern shore between Msalaba and Luagala Point we encounter *P.* sp. "orthognathus ikola". There is little geographical variation within each range, and hence I consider these three forms to be separate species. Females of the *P. polyodon*-like species have a subdued colour pattern consisting mainly of a light-coloured body and vertical bars; females of the *P. orthognathus* group, however, have a colour pattern which is virtually identical to that of males. Members of both species-groups occur in the same habitat but it seems that *P. orthognathus* types are more tolerant of sediment-covered rocks.

Petrochromis famula

A *Petrochromis* species which shows considerable geographical variation and which has formerly been erroneously identified as *P. polyodon*, is now known to be *P. famula* (Herrmann, 1996). It occurs on almost all rocky shores, with the type material originating from the northern part of the lake. The total length of *P. famula* is noticeably less than that of the previous two species and males attain about 15 cm maximum. Although *P. famula* shares the habitat with larger herbivorous species, males seem to be able to secure territories, mostly occupying caves that are too small for *P. polyodon* and *P. ephippium*. Females and non-territorial males sometimes group together into large foraging schools and move through the habitat feeding at any site they choose (see photo page 45). Because of their large numbers they can enter any territory — no tenant can hope to chase away several hundred fishes! This method of evading aggressive attacks from territory-owners is observed

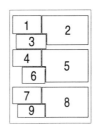

page 48

1. *Petrochromis* sp. "texas red" (Ubwari, Congo) in the aquarium.
A male (2) and female (3) *Petrochromis* sp. "kasumbe" at Halembe, Tanzania.
4. *Petrochromis* sp. "texas kipili" at Msalaba, Tanzania.
A male (5) and female (6) *Petrochromis* sp. "texas" at Mabilibili, Tanzania.
7. *Petrochromis polyodon* at Kantalamba, Tanzania.
A male (8) and female (9) *P. polyodon* at M'toto, Congo.

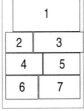

page 49

1. *Petrochromis* sp. "orthognathus ikola" at Msalaba, Tanzania.
A male (2) and a mouthbrooding female (3) *P.* sp. "orthognathus ikola" at Kekese, Tanzania.
4. *Petrochromis* sp. "orthognathus tricolor" at Chimba, Zambia.
5. *P.* sp. "orthognathus tricolor" at Mvuna Island, Tanzania.
6. *Petrochromis* cf. *orthognathus* at Cape Tembwe, Congo.
7. *Petrochromis orthognathus* at Magara, Burundi.

in other species as well (e.g. *P. fasciolatus* and *Tropheus moorii*).

Petrochromis ephippium and its allies

Petrochromis ephippium and *P. trewavasae* resemble each other and have sometimes been regarded as conspecific (e.g. Kuwamura, 1987b), but at several locations along the southern shores of the Congo (e.g. Kanoni and Kapampa) I found both species sympatrically. *P. ephippium* has a lakewide distribution whereas *P. trewavasae* is restricted to the southwestern part of the lake, with the highest density of individuals at Kapampa. *P. trewavasae* is an extremely shy species and is usually found in small groups. Both species can attain a total length of about 20 cm. *P. trewavasae* does not exhibit geographical variation, and both male and female have an almost identical colour pattern consisting of a brown to black ground colour with many light yellow-coloured spots. A similar pattern is also seen in *P. ephippium*, but this species has an additional light-coloured patch along its back and in the dorsal fin (the "saddle-spot"). The colour of this saddle-spot is either white or yellow, depending on the locality. There are both white- and yellow-saddled populations in the southern part of the lake, but the marking is yellow in those of the central eastern coasts and white in those from the northern half of the lake. I have seen large, mouthbrooding, females which exhibited the black ground colour with light-coloured spots, but I have seen some (not all at the same site) territorial males which have lost this pattern and turned to a dirty yellow (in the south, e.g. Mvuna Island; see photo page 52) or gray-blue (in the north, e.g. Halembe).

There is another species (at least I think it is different from *P. ephippium*) in which adult males are dirty to bright yellow, and which is known as *P.* sp. "moshi yellow". It occurs along the eastern coast between Msalaba and Luagala Point, and between Cape Kabogo and Kigoma. The males at Maswa have a yellow body with a dark brown head, but those at Kigoma are all yellow. The females of this bright yellow species resemble *P. ephippium* in many respects, although they do not have the light-coloured spots. They do, however, have a cream-coloured saddle and a dark grey-brown body. I am still not certain whether the "Moshi Yellow" is different from full-grown *P. ephippium*, as some males of the latter resemble

discoloured "Moshi" males. At Kekese I was able to take photos of both forms/species. Here *P. ephippium* has a sulphur-yellow saddle, whereas this marking is white in females of the "Moshi Yellow" (see photos on page 52). Still, the yellow-saddled *P. ephippium* could nevertheless be sub-adult males, but it would be uncommon to find juvenile males with a colour pattern different to that of females. And if you think this is complicated what about the following few *Petrochromis* species which are found sympatrically with the previously mentioned *Petrochromis*?

Rare Petrochromis

P. sp. "kipili brown", which occurs in the rocky habitat between Fulwe Rocks (near Wampembe) and Nkondwe Island (near Kipili), is an interesting species. Females have a heavily barred pattern, but males are mostly chocolate brown. Males of the populations near Kipili have yellow bars on the lower half of the body (in fact these are the yellow ground colour, visible because of the narrower vertical bars) whereas those at Fulwe Rocks lack such bars, but have an orange edge to the tail. The "Kipili Brown" is found in shallow water and also as deep as 20 metres, which is remarkable for a herbivore.

Near Kipili this species is sympatric with the "Texas Kipili" and with a third *P. polyodon*-like species: *P.* sp. "texas blue". I have seen the "Texas Blue" only south of Kipili near Kisambala, in rather shallow water. I could not distinguish females as the males interacted with all other *Petrochromis* species and were rather shy.

The "Red Petrochromis" hails from Luagala Point (Horst Dieckhoff, pers. comm.) where it is found in rather deep water, deeper than 15 metres. It is possible that *Petrochromis* sp. "red" is a geographical variant of the "Moshi Yellow", because the latter tends to be more orange-coloured along the Mahali mountain range, but since I have not personally been able to observe this form in the lake, I will treat it here as a separate species.

In captivity all *Petrochromis* species are problematic. The males often harass females at every opportunity. A large group of females and one male does not seem to solve the problem. Some hobbyists have succeeded with a single pair in a large tank with plenty of refuges. If a large tank is not available then a surplus of males (not females) is recommended. Another problem is their diet. As in many herbivores, *Petrochromis* may develop digestive or other physiological disorders if fed large quantities of protein-rich or otherwise unsuitable food (such as beef heart or red mosquito larvae), and need to be fed top quality vegetable-based flake or pelleted food. *Petrochromis* species are in no respect good aquarium residents, and need to be left to the experienced aquarist.

Tropheus species

The genus *Tropheus* is currently composed of eight species: *T. moorii* (*T. kasabae* is a junior synonym), *T. annectens* (*T. polli* is a junior synonym), *T. duboisi*, *T. brichardi*, *T.* sp. "black", *T.* sp. "red", *T.* sp. "mpimbwe", and *T.* sp. "ikola". The taxonomy of this genus is confusing, not least because no-one has seen every population or form in its natural habitat. As far as I know there is no rocky coast in the lake which does not harbour a *Tropheus* species, and at least four of the above-mentioned species show great geographical variability, confusing the issue even further. Fortunately most aquarists are aware of this fact and rarely house different variants of the same species in the same tank, as this would inevitably lead to hybridization.

These funny-looking cichlids are very popular among aquarists. Unfortunately almost all *Tropheus* species are referred to in the aquarium hobby as *T. moorii*, which has confused quite a few hobbyists as well as scientists. *T. moorii*, the type species of the genus, has a restricted distribution in the southern part of the lake. Its northern counterpart, *T.* sp. "black", exhibits very similar behaviour, but at a small rocky outcrop along the central coast of Tanzania both species can be found sympatrically, indicating that these really are two different species.

Not only for their protection, but also for their food, *Tropheus* species are restricted to the upper rocky habitat. All *Tropheus* species are algae browsers which shear and tear the algal strands from the rocks and avoid the ingestion of inorganic sediment (Yamaoka, 1983a). The outer teeth of *Tropheus* species are bicuspid and set closely together forming a tight-packed row with which the fish can grasp individual algal strands, which are severed from the substrate by biting or, most often, by swinging the head sideways while holding tight to the strands.

Takamura (1984) and other investigators have found that *T.* sp. "black" has a feeding relationship with several species of the genus *Petrochromis*. The latter are algae grazers which mainly comb loose material, including some sediment, from the biocover. Sites grazed by these *Petrochromis* species contain less sediment and are favoured by *Tropheus* species (I have seen this behaviour in southern *Tropheus* species as well). It is also interesting to observe that on many occasions a *Tropheus* species is not chased from the feeding territory of a large and aggressive *Petrochromis*, but allowed to feed on the already-grazed areas. Such sediment-free browsing sites are limited, however, and not infrequently one can observe large foraging schools, numbering hundreds of individuals of a single *Tropheus* species, roaming through the rocky habitat and feeding at preferred sites. Such

1	6
2	7
3	8
4 5	9

page 52

1. *Petrochromis ephippium* at Sibwesa, Tanzania.
2. *P. ephippium* at Ulwile Island, Tanzania.
3. *P. ephippium* at Kantalamba, Tanzania.
4. *P. ephippium* at Halembe, Tanzania.
5. *P. ephippium* at Mvuna Island, Tanzania.
A male (6) and female (7) *Petrochromis* sp. "moshi yellow" at Kekese, Tanzania.
8. *Petrochromis trewavasae* at Kapampa, Congo.
9. *P. trewavasae* at Moliro, Congo.

1	5
2 6	7
3	8
4	9

page 53

1. *Petrochromis* sp. "texas blue" at Kisambala, Tanzania.
2. *Petrochromis* sp. "red" (from Luagala Point, Tanzania) in the aquarium.
3. *Petrochromis famula* at Kapemba, Zambia.
4. *Petrochromis famula* at Cape Tembwe, Congo.
5. *Petrochromis* sp. "kipili brown" at Kerenge Island, Tanzania.
6. A female *P.* sp. "kipili brown" at Mvuna Island, Tanzania.
7. *P.* sp. "kipili brown" at Fulwe Rocks, Tanzania.
8. *P. famula* at Kambwimba, Tanzania.
9. *P. famula* at Chimba, Zambia.

sites are too heavily defended to be accessible to the individual, but force of numbers permits them to be browsed. I have seen foraging schools formed by *T. moorii, T. brichardi, T. annectens, T.* sp. "red", and by *T.* sp. "black".

All *Tropheus* species are territorial and most males have rather large feeding territories, sometimes with a diameter of more than two metres (e.g. in *T. brichardi*). Conspecific males, and often — but not always — females too, are chased from such territories. Interestingly, females also have feeding territories from which conspecific females are chased (Kuwamura, 1992; 1997). These territories are neither very large nor aggressively defended, as it is not uncommon to see a few individuals feeding from the same small rock.

All *Tropheus* species are maternal mouthbrooders, but only a couple of species (*T. annectens* and *T. brichardi*) exhibit the sexual dichromatism commonly observed in maternal mouthbrooders. Kawanabe (1981) found that a female *T.* sp. "black" is allowed in the male's territory for up to 3 weeks prior to spawning, but that the female leaves the male's territory after spawning and hides among the rocks. Female *T. duboisi*, however, return to their original feeding territory after spawning (Yanagisawa & Sato, 1990). I have seen territorial male *T. moorii* and *T.* sp. "red" chasing some conspecifics from their premises but ignoring others, suggesting that these two species also form a kind of relationship between male and female before spawning (as in *T.* sp. "black"). Yanagisawa and Nishida (1991) speculate that the female's oocytes can ripen only when she is under the protection of a territorial male.

Mouthbrooding females forage carefully from the biocover in order to feed the fry inside their mouths and probably also themselves, although they appear very emaciated after four weeks of mouthbrooding. Females of *T. duboisi, T. moorii, T.* sp. "red", and *T.* sp. "black" guard their offspring for a few days after they have been released for the first time. In the aquarium it is *T. duboisi* that shows the longest care; usually for about a week. In the confines of the aquarium females may not guard their offspring at all. *T. duboisi* juveniles are released in the female's territory, but those of the other species are commonly found in the extreme shallows among pebbles and small rocks. In the aquarium too, fry of *T. moorii, T. brichardi*, and *T.* sp. "black" are released in the uppermost rocky area (Tijsseling, 1982).

Tropheus moorii

The type specimen of *T. moorii* was caught at Kinyamkolo, the old name of Mpulungu. The *T. moorii* found at Mpulungu (actually Kasakalawe, which is a few kilometres west of the town) is called "Sunset Moorii" in the trade and is characterized by an orange blotch on the flanks. Nelissen (1977) described the subspecies *T. moorii kasabae* using specimens caught at Cape Chaitika. The trade name of this form is "Blue Rainbow Moorii" and it was elevated to full specific rank, as *T. kasabae*, by Poll (1986). *T. kasabae* is, however, here regarded as a junior synonym of *T. moorii*. It is unclear from Nelissen's paper whether he compared the "Blue Rainbow Moorii" with *T. moorii* from Mpulungu or with specimens of the northern *T.* sp. "black", as the latter were common in the aquarium trade, and regarded as *T. moorii*, at the time of publication. If he used *T.* sp. "black" from Burundi then he would have found not only a difference in colour and distribution between the northern species and *T. moorii kasabae*, but also a morphometric distinction.

Most geographical variants of *T. moorii* found in Tanzania are rarely imported, except for the beautiful "Red Rainbow Moorii" from Kasanga, which is rather common along the rocky coast at Kasanga (Bismarck Point) and further south. The site where almost all "Red Rainbows" are collected for the aquarium trade lies a few kilometres south of Kasanga. There is some variation in the *T. moorii* population between Kasanga and the Kalambo river, the border with Zambia: although there is a single continuous population, some individuals exhibit more blue pigment in the dorsal fin than others. Most, however, have a distinctly red dorsal. In my opinion the blue colour in the dorsal does not imply two different populations but is instead simply part of individual variation (such as dark or blond hair in humans).

At Kasanga a sandy bay about 4 km long separates the "Red Rainbow" population from the so-called "Blue Blaze Moorii". The "Blue Blaze" — the trade name is misleading as it does not have a blue blaze but blue spots in the dorsal — lacks the bright yellow blotch seen on the body of the "Red Rainbow", and has more blue spots in the dorsal fin. The females show a little of the yellow colour, especially on the lower half of the body, but its intensity is much reduced compared to that of the "Red Rainbow". The distribution of the "Blue

Blaze" is restricted to a few rocky headlands near the village of Muzi, north of Kasanga.

At Samazi, about 12 km north of Kasanga, *T. moorii* lacks any bright colours: the red and blue in the dorsal are much reduced compared to the "Blue Blaze Moorii". At Katili *T. moorii* exhibits a brown-green coloration and the dorsal fin lacks the red pigment almost completely. The head and chest are still red, but this population is probably the least attractive.

At Mamalesa Island (Malasa) some *T. moorii* have a few small, vertically elongated, yellow spots on the flank. The dorsal is yellow and there is a yellowish-grey hue all over the body. *T. moorii* at Kala Island has yellowish fins as the most prominent feature of its coloration. The two islands at Kala lie close to shore and it is very unlikely that they would harbour a *T. moorii* population different to that found at the mainland, which has not been visited.

Further north the coastline differs from that south of Kala. Even though there are high mountains close to the lake, over the millennia so much sediment has been washed down by the rivers that very flat — and probably fertile — shores have been formed. These flats are more than 1 km wide in some places, and have buried the rocky habitat over a long stretch between Kala and Wampembe. Approximately 1 km offshore at Wampembe there is a small reef formed by huge boulders. This reef, known locally as Fulwe, barely breaks the surface and lies at a depth of approximately 25 metres. At this locality *T. moorii* has a yellowish brown colour which is not at all unattractive. Further north the structure of the shoreline is similar to that south of Wampembe: sandy beaches without rocky outcrops.

At Hinde B the rocks are very large, and the sand, at about 12 metres of depth, very coarse. *T. moorii* at Hinde B closely resemble the variant at Fulwe but the narrow bars are more yellow and the fish is more attractive in appearance. The variant of *T. moorii* seen at Hinde B differs from other known populations in that even adult male specimens exhibit the narrow vertical barring normally seen only in juveniles (of all forms). Adult females have prominent barring, the bars being narrow and pale yellow in colour.

Between Hinde B and Namansi the shoreline consists of sandy bays alternating with small rocky outcrops. The coloration of the *T. moorii* found between these two localities changes slightly from brown (Hinde B) to yellowish brown (Namansi) and the first three vertical bars behind the gill cover become more prominent and slightly widened.

Between Namansi and Kisambala *T. moorii* and *T. brichardi* are both found sympatrically. *T. moorii* is found in the upper part of the rocky habitat where there is hardly any sediment, sometimes in large numbers in foraging schools. By contrast *T. brichardi* occurs in small numbers at depths of more than 10 metres, sometimes down to 28 metres, where a visible layer of sediment covers the rocks. The individuals of *T. brichardi* are much smaller than those of *T. moorii* and are also smaller than *T. brichardi* found in nearby areas where *T. moorii* is absent. Although it seems that *T. brichardi* is partially tolerant of sediment-covered feeding grounds (usually found in deeper areas), it probably does not find sufficient food to give it the size and strength to compete with *T. moorii* in obtaining a territory in the shallower areas. *T. moorii* is noticeably larger, and thus stronger, than *T. brichardi* in this area, and is able to hold territories in the "better" parts of the rocky habitat.

Of great interest is the fact that *T. moorii* shows a large gap in its distribution along the eastern shore of the lake: between Kisambala and Sibwesa, a stretch of coastline of about 200 km, it is absent, but a relict population occurs at Lyamembe, a few kilometres north of Sibwesa. This population of *T. moorii* is sympatric with *T.* sp. "black", the so-called "Kirschfleck Moorii", and with *T. annectens* (previously erroneously known as *Tropheus polli* — see below).

The distribution of *T. moorii* along the Zambian and Congolese shores also exhibits a gap, specifically between Nkamba Bay, Zambia, and Kiku, Congo. In this gap we find *T.* sp. "red" (see below), a species closely related to, and very likely

page 56	Geographical variants of *Tropheus moorii* and their ranges along the western side of the lake.
page 57	*T. moorii* variants and their ranges along the eastern lake shore. Note the isolated distribution of a small population near Sibwesa (Lyamembe) in the central part of the lake (top of page).

① Kanoni

② Lusingu

③ Tembwe

④ Kapampa

⑤ Nkamba Bay

⑥ Cape Nangu

⑦ Lufubu

⑧ Cape Chaitika

⑨ Katoto

⑩ Kasakalawe

⑪ Mbete Island

⑫ Chituta

① Lyamembe

② Kisambala

③ Mtosi

④ Hinde B

⑤ Fulwe

⑥ Mamalesa Is.

⑦ Katili

⑧ Samazi

⑨ Muzi

⑩ Kambwimba

⑪ Isanga

derived from, *T. moorii*. I regard the red-coloured *Tropheus* as different from *T. moorii* because several independent observers have seen both species sympatrically at Kiku and near the Kasaba Bay Lodge in Nkamba Bay (Brichard, 1989; Toby Veall, pers. comm., Christian Houllier, pers. comm.).

South of the Kalambo River, which forms the border between Zambia and Tanzania, along the eastern side of Chituta Bay, we find the red-eyed "Chisanza Moorii". This *Tropheus*, as well as being basically a very attractive variant, exhibits socalled OB (orange-blotch) polychromatism, which makes some specimens extremely colourful. Such orange-blotched coloration, well-known from many of the mbuna (rock-dwelling cichlids) of Lake Malawi, is very rare in Lake Tanganyika and has been found only in *Ctenochromis benthicola* (only orange individuals, no blotched forms) and *T. moorii*. I have seen blotched *T. brichardi* but these didn't have an orange ground colour. The OB forms of *T. moorii* are known as "Golden Kalambo Moorii" in the hobby and the special coloration occurs only in fully adult individuals. As well as the true OB specimens, a larger population of forms with yellow edges to the dorsal and caudal fins is found in the same area. Probably some of these are young adults which will develop the true OB pattern when mature, but others exhibit the yellow-edged pattern when fully adult. The rocky habitats along the eastern side of the Chituta Bay are rather small and so are the populations of *T. moorii* inhabiting them. Small, and possibly vanishing, populations may have led to the appearance of these OB morphs, as more individuals are inbreeding than they would in large populations.

The *T. moorii* found between Kasanga and Cape Chaitika are characterized by blue spots on a light-coloured head. The two forms found between the Lufubu River and Nkamba Bay, the socalled "Nangu Moorii" and the "Ilangi Moorii", lack such spots but have a yellow-coloured body which is one of the distinguishing characteristics between *T. moorii* and *T.* sp. "red". The "Ilangi Moorii", found south of the Kasaba Bay Lodge in Nkamba Bay, occurs sympatrically with a darkred form of *T.* sp. "red", the so-called "Chilanga Moorii" (Veall, pers. comm.).

At Kiku, where only a small population of *T. moorii* is present (Houllier, pers. comm.), the most common *Tropheus* is *T.* sp. "red", characterized by a blood-red colour on throat and breast. The *T. moorii* at Kiku resembles that of Kapampa. Continuing northward along the Congolese shore, *T. moorii* occurs as far north as Moba. One of the most northern populations in the Congo is the socalled "Murago Moorii", which is found south of Moba at a place called Lusingu.

Tropheus sp. *"red"*

A few kilometres north of the Lunangwa River, at Kiku, a red-coloured variant of *Tropheus* is found in the upper rocky habitat. This population differs noticeably from the one at Kapampa, but no clear boundary between these two populations can be established. Christian Houllier (pers. comm.) also found the "Kapampa Moorii" at Kiku, where it lives sympatrically with the red variant. This indicates that the "Red Moorii" is a true species and distinct from *T. moorii*. Brichard (1989) reports the sympatry of a variant of the "Red Moorii" (his Nsumbu-Nkamba line) and a variant of *T. moorii* (Kabeyeye line) in Nkamba Bay. Again this would indicate that the "Red Moorii" is a different species.

The characteristics of the red *Tropheus* are a dark body with a bordeaux-red hue, two chevron-like stripes between the eyes, and sometimes a brightcoloured stripe on the lower cheek, in particular in males. Another difference between *T. moorii* and *T.* sp. "red" (and most other *Tropheus*) is the blue colour of the first ray in the pelvic fins — this characteristic is found in all populations of *T. moorii*, including the one near Sibwesa. The anterior edge of the pelvic fins in *T.* sp. "red" is dark brown or reddish, but not blue. Underwater, *T.* sp. "red" is readily identified as most individuals show a dark to almost black coloration. None of the populations known have light-coloured flanks, the most obvious characteristic of *T. moorii*. Territorial males displaying aggressive behaviour exhibit the coloured stripe on the cheeks.

The distribution of the "Red Tropheus" fills the gap in the distribution of *T. moorii* with overlap at either end of the range. These overlapping areas are rather small, several 100 metres, and comparable to that of *T. moorii* and *T.* sp. "black" at Lyamembe (Sibwesa), but unlike the 10 km overlap of *T. moorii* and *T. brichardi* south of Kipili. The most common *T.* sp. "red" in the hobby are the socalled "Tailstripe Moorii" from Cape Kipimbi (Chipimbe) and Katete in Zambia, and the form from Cape Kachese, also in Zambia.

South of Kiku, near Livua, the so-called "Blood-throat Moorii" inhabits the rocky habitat; a beautiful variant which has only sporadically been exported for the aquarium trade. Between Moliro, Congo, and Katete, Zambia, *T.* sp. "red" is characterized by a red triangle on the posterior part of the body including the caudal peduncle. The stripe on the caudal peduncle can vary from a broad blood-red band to a complete absence, via a few dots, and this can be seen within a single population. In most populations of *Tropheus* we find a noteworthy variation in colour pattern pervading any particular population, but this is just a sign of the variability of the genus and of cichlids in general.

The "Tailstripe Moorii" closely resembles the next distinguishable population, at Chimba in Cameron Bay. The Chimba variant has in the past been imported from Zambia as the "Red Moorii". The individuals found north of Ndole Bay are a little less red and show a more orange colour on the sides. The coast is sandy but just at the shoreline there is a small rocky zone formed by sandstone. The *Tropheus* living at less than one metre of depth can be regarded as an intermediate population between the Chimba and the Kachese forms. *T.* sp. "red" from Cape Kachese has an orange-coloured cheek and an orange-red dorsal fin. The individuals in the populations at Sumbu Island, Chisanse, and Chilanga (Nkamba Bay) have a general dark-red body and a red dorsal fin. The coloured stripe on the cheek is less prominent in these forms.

Tropheus sp. "black"

T. sp. "black", from the northern half of the lake, also shows geographical variation. The basic colour is, however, black or dark brown. The population at Uvira is completely black, and between Uvira and Luhanga the population consists of fishes which have an overall black colour when adult (sometimes with yellow patches), but have an orange bar on the flanks when young (Brichard, 1989). At Pemba (previously Bemba) a very attractive form of *T.* sp. "black" inhabits the rocky habitat, together with *T. duboisi*. The so-called "Bemba Moorii" is black with a very wide bright orange bar on the body. At Mboko there are three shallow islands which are inhabited by the "Lemon Stripe Moorii" or "Mboko Moorii",

which exhibits a very narrow yellow bar on a black body. Two sediment-rich rocky outcrops near Lueba harbour another variant. According to Brichard (1989) the population found 3 km north of Lueba is very variable, ranging from entirely black morphs to those with one or two bright yellow patches on the flank. The fish of the population at Lueba are entirely black.

The very popular "Kiriza Moorii" or "Kaiser II Moorii" hails from the western shores of the Ubwari peninsula and is separated from the completely differently coloured "Caramba Moorii" by a variant which shows a variable yellow patch or bar on a black body ("Banza Moorii"). According to Brichard (1989), who thoroughly investigated the northern region of the lake, the range of the "Caramba Moorii" lies between Cape Caramba, at the southern tip of the peninsula, and Cape Muzimu, almost at the northern tip. This would mean that there is no obvious physical barrier between the distribution of the "Caramba Moorii" and that of the "Banza Moorii", which is found between Cape Muzimu and Manga, a village at the northernmost tip of the peninsula. The "Caramba Moorii" shares the habitat with another *Tropheus* species: *T. brichardi*, the so-called "Ubwari Green Moorii" (Brichard, 1989).

The northernmost part of the lake consists of sandy beaches and is completely devoid of rocky areas. The most northerly rocky shores on the eastern side of the lake are found near Magara in the Rutunga district of Burundi. This area is inhabited by the so-called "Brabant Moorii", the very first *Tropheus* ever to appear in the aquarium trade (in 1958). South of Magara, at Minago, *T.* sp. "black" has a black overall colour and further south, at Rumonge, it is again black, but with a faint brown-yellow blotch on the belly.

There is a considerable gap in the distribution of *T.* sp. "black" along the eastern shore of the lake. The northeastern populations are found between Magara and Rumonge and the central eastern

| page 60 | Geographical variants of *Tropheus* sp. "red" and their ranges along the southwestern shore of the lake. |
| page 61 | *Tropheus* sp. "black" variants and their ranges in the northern section of the lake. |

① Kiku

② Livua

③ Cape Kipimbi

④ Katete

⑤ Chimba

⑥ Cape Kachese

⑦ Sumbu Is.

⑧ Chisanse

⑨ Chilanga

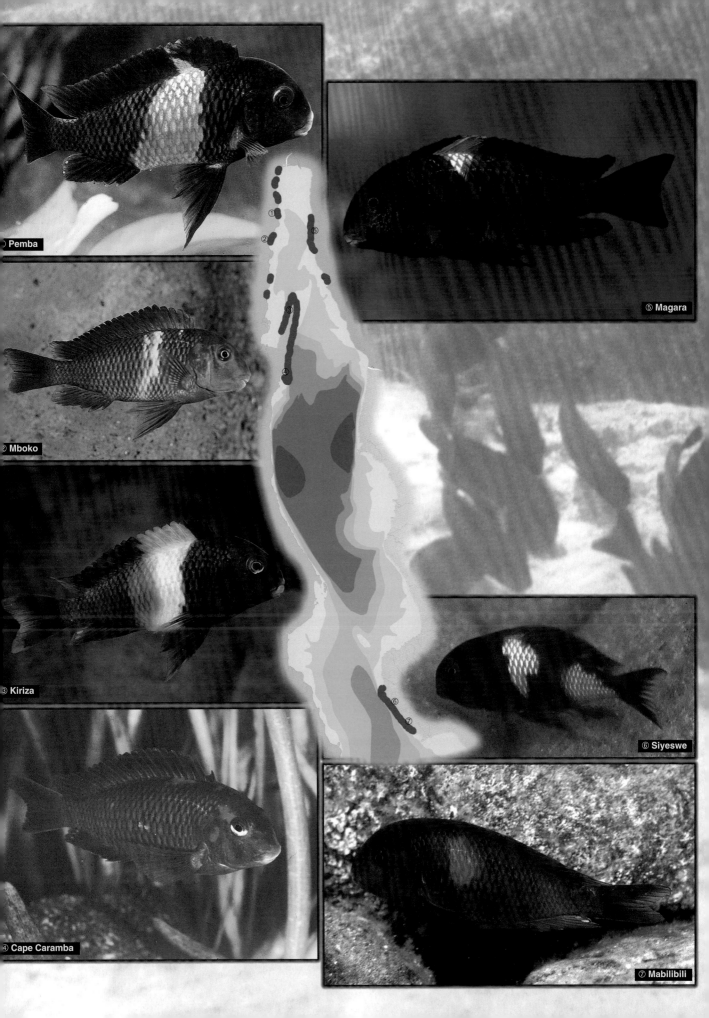

① Pemba

② Mboko

③ Kiriza

④ Cape Caramba

⑤ Magara

⑥ Siyeswe

⑦ Mabilibili

populations between Luagala (also Luahagala) and Lyamembe (near Sibwesa). The latter populations also include the popular "Kirschfleck Moorii". The entire shoreline of the central-eastern range is rocky apart from very small sandy bays. There seems to be a cline in the coloration of *T*. sp. "black" along this part of its range. The most southerly populations at Lyamembe and Mabili-bili are black with two dark-red blotches on the flank; sometimes the blotches are combined to form an almost dark red flank. Lumbye Bay and Siyeswe Bay, and the areas in between, harbour the "Kirschfleck Moorii" that is exported for the aquarium trade, and here they have the most attractive colour. Further north, towards Lubugwe Bay and around Luagala Point, *T*. sp. "black" is mostly black with very little red colour. Over its entire central distribution *T*. sp. "black" shares the habitat with *T. annectens* (previously known as *T. polli*) and at Lyamembe both species are found sympatrically with *T. moorii*. Earlier reports (Konings, 1988) that the "Kirschfleck Moorii" was also found at Karilani Island and Bulu Point were based on personal communication with Horst Dieckhoff, but could not be confirmed after visiting these localities.

Tropheus brichardi

T. brichardi was described from specimens collected near Nyanza Lac in Burundi (Nelissen & Thys van den Audenaerde, 1975). This population is known as the "Chocolate Moorii". The main characteristic of *T. brichardi*, which seems to have the widest distribution of all *Tropheus* species, is its colour pattern of distinct vertical bars, which is also visible in adult females. Most large adult males, however, lose this pattern or have just residual traces of it on a generally green or brown (but never black) body.

It appears that *T. brichardi* tolerates a higher level of sedimentation on the substrate and can therefore be found in silted areas as well as in deeper rocky regions. This may also have facilitated its dispersal along both the eastern and western shores of the lake. In the aquarium *T. brichardi* appears to be more aggressive and determined in its territorial defence than other *Tropheus* species, and this determination is also noticeable in its natural setting. Interestingly, where it occurs sympatrically with *T. moorii*, south of Kipili, it occupies the less-preferred part of the rocky habitat: here it is found in regions deeper than

10 metres whereas it is found in the upper regions at places where *T. moorii* is absent. Although I believe that the aggression level of the individuals of a species is an important factor in securing territory, in this case it may be that *T. brichardi* is subordinate to the perhaps better feeding technique of *T. moorii*. Another scenario could be that *T. moorii* is tolerated by the larger and more aggressive *Petrochromis* species dominating the upper rocky habitat whereas *T. brichardi* is not. *T. moorii* feeds only from sediment-free sites but *T. brichardi* may also feed from sites that are preferred by *Petrochromis*.

I have not visited the Ubwari Peninsula, where *T. brichardi* is found sympatrically with *T*. sp. "black", but a similar arrangement may exist here as well: *T*. sp. "black" in the upper, algae-richer, rocky habitat and *T. brichardi* in the sediment-richer deeper regions. Brichard (1989) reports that a *T. brichardi*-like species is found sympatrically with *T*. sp. "black" at Uvira but that *T. brichardi* is strongly outnumbered by the other. This may suggest that *T*. sp. "black" at Uvira is also the better adapted species of the two.

From the border between Burundi and Tanzania to Kigoma the rocky habitat is inhabited by the "Kigoma Moorii". This variant of *T. brichardi* is dark-green to blue-brown when adult and very common among the rocks. South of Kigoma, at Ujiji, the so-called "Katonga Moorii" is collected for the aquarium trade. Adult males of this form always exhibit a barred pattern. South of the Malagarasi River, as far as Halembe, *T. brichardi* is generally green-brown with only a population at Maswa having a different colour; here the so-called "Green Wimple Moorii" has a green body with a broad yellow bar. Between Halembe and Bulu Point there is no rocky coast, but at Bulu Point, Karilani Island, and along the mainland further south to the Kasoje River outlet, the shoreline harbours a dull brown-coloured variant of *T. brichardi*. This population is sympatric with *T. annectens* between Kasoje River and Bulu Point.

Between Kasoje River and Cape Korongwe, north of Kipili, there is no *T. brichardi* although two *Tropheus* forms found in this area were previously assigned to this species. These two forms, *T*. sp. "ikola" and *T*. sp. "mpimbwe", are now regarded as being different from *T. brichardi* on grounds explained later (page 63). Between Cape Korongwe (near the village of Kapemba) and Kisambala, south of Kipili, *T. brichardi* is the only member of its genus found. The forms caught at the islands near Kipili have been exported as "Kipili Moorii" and are characterized by attractively-coloured juveniles. Between Kisambala and Namansi *T. brichardi* shares the habitat with *T. moorii* and is found at much deeper levels of the rocky biotope than usual. At Mtosi I have seen *T. brichardi* at a depth of 28 metres which is rather deep for an algae-feeding cichlid. Dieckhoff (pers. comm.) photographed an attractive *T. brichardi*-like cichlid, with a bright yellow colour at the base of the pectoral fins, "at Isonga", a village in the mountain range north of Karema, Tanzania. The shores in this area are entirely rocky, but I found only *T. annectens* and *T.* sp. "ikola" here. I suspect that the "Isonga Moorii" may have been photographed at a different locality, possibly close to the Malagarasi River delta.

T. brichardi also occurs along the western shores of the lake, and is found between Uvira (Kavimvira: Brichard, 1989) and Kambwebwe, south of Cape Tembwe. The "Green Ubwari Moorii", found along the eastern shores of the Ubwari Peninsula, has not, or only rarely, been exported for the aquarium trade. The shores between the Kavala Islands and Cape Caramba are said to harbour two different *T. brichardi* types (Brichard, 1989), but although I have dived at two of the Kavala islands I was unable to find more than a single *T. brichardi* species. The descriptions given by Brichard of the two species do not differ very much and it would be interesting to observe these two species side by side in their natural habitat. At Kalemie and further south, as far as M'toto, *T. brichardi* is chocolate brown. At M'toto it is sympatric with *T. annectens*.

Tropheus sp. *"mpimbwe"*

T. sp. "mpimbwe" has previously been assigned to *T. brichardi*, but because it is found in an area of the lake which is relatively new, and because its juveniles differ dramatically in coloration from neighbouring *Tropheus* populations, I prefer to treat it as a separate species. It occurs at Cape Mpimbwe (also Msalaba) and does not overlap, as far as is known, with *T. brichardi*. I found *T. brichardi* (the "Kipili Moorii") at Cape Korongwe, near the village of Kapemba, and recently Karlsson (1998) reported the yellow-cheeked form in Korongwe Bay. Thus the two species, *T. brichardi* and *T.* sp. "mpimbwe", appear to be separated by a sandy shore with a length of approximately 2 km. The orange-cheeked form of *T.* sp. "mpimbwe" occurs only a little south of the village of Mpimbwe (Karlsson, 1998). A yellow-cheeked form is further found north at the cape.

Juvenile *T.* sp. "mpimbwe" are not orange or bright yellow like those of *T. brichardi* at Kapemba or Kipili; they have an attractive pattern of dark brown bars on a white or pale yellow background. This pattern closely resembles that of *T. annectens* juveniles and at some time in the past the red-cheeked "Mpimbwe Moorii" was exported as "Red Polli". However, *T. annectens* is characterized by 4 spines in the anal fin whereas all other *Tropheus* species, including *T.* sp. "mpimbwe", have 5 to 7.

At Kalila, north of Kabwe and south of Karema, I found a *Tropheus* species which does not resemble the Mpimbwe Moorii, the "Kaiser Moorii" (*T.* sp. "ikola" — see below), nor the Kipili Moorii. On closer inspection of sub-adults we may find that these resemble *T. brichardi* types more than those of the Kaiser Moorii. Although I have previously assigned the form at Kalila to *Tropheus* sp. "ikola", I now believe that it is closer to *T.* sp. "mpimbwe".

The large rocks at Kabwe, about 10 km north of Cape Mpimbwe, lie in very shallow water but surprisingly harbour a small population of *Tropheus*. The males of this population do not have a coloured patch on the cheek (like *T.* sp. "mpimbwe") or a broad yellow bar (like the "Kaiser Moorii"), but further details of their colour pattern cannot be given owing to the fact that visibility at this exposed site is bad almost all year round. This population may also be part of *T.* sp. "mpimbwe".

Tropheus sp. *"ikola"*

This species is better known as the Kaiser Moorii (Kaiser is German for emperor) and is very popular among aquarists. The Kaiser Moorii is restricted to the rocky shores between Ikola and Isonga. The shore at Ikola is pure sand and the rocky coast starts about 20 km north of the village. At the northern end of the rocky area, near Isonga, the shore projects at an almost 90 degrees angle into the lake. The entire shoreline as far as

page 64	Geographical variants of *Tropheus brichardi* and their ranges in the northern part of the lake.
page 65	*Tropheus brichardi* variants and their ranges around Kipili in Tanzania. The small photos show the juveniles of five different populations.

① Kabimba

② Kavala

③ Cape Tembwe

④ Kambwebwe

⑤ Nyanza Lac

⑥ Katonga

⑦ Maswa

⑧ Halembe

⑨ Bulu Point

⑩ Karilani Is.

① Kapemba

② Nkondwe Is.

③ Kerenge Is.

④ Mvuna Is.

⑤ Ulwile Is.

⑥ Mkinga

⑦ Mtosi

① Kapemba

② Nkondwe Is.

③ Kerenge Is.

⑤ Ulwile Is.

⑧ Kipili

Sibwesa, about 18 km, is sandy. *T*. sp. "ikola" seems to live in a single, continuous population and geographical variants are not known.

Tropheus annectens

The members of this species can readily be identified because they are characterized by the possession of four spines in the anal fin whereas all other *Tropheus* species have 5 to 7 anal spines. A further characteristic is that adult males have a colour pattern noticeably different from females, and, together with *T. brichardi* and *T*. sp. "mpimbwe", form a small group within the *Tropheus* genus which show sexual dichromatism. The holotype of *T. annectens* was reportedly collected at Kalemie (Boulenger, 1900), but it is likely that the collector stayed at Kalemie rather than that he collected the fish there. *T. annectens* occurs along the western shoreline only near M'toto, in an area not much longer than 15 to 25 km. It is, however, also found on the opposite shore between Bulu Point and Ikola. Like all the previously mentioned *Tropheus* species it prefers the upper rocky habitat and is rarely found deeper than 7 metres. Even though there are large stretches of sandy shore (from which *Tropheus* is absent) in the eastern distribution of *T. annectens*, the three to four rather isolated populations do not show any significant geographical variation. And, remarkably, the population on the opposite coast at M'toto is not much different either. Nevertheless the latter population must have been isolated from the others for a very long period, at least since the time the water level was much lower and a shallow rocky habitat was present between the eastern and western shore. Perhaps because of the species' sympatry with *T. duboisi*, *T. brichardi*, and *T. moorii* at various places in its range, female *T. annectens* may have become very critical in selecting their mates, with this putative stringent sexual selection resulting in very little variation among the *T. annectens* populations.

The forms along the eastern shore have been described as *T. polli* and are still well-known among aquarists under this name. There is little doubt that *T. polli* is a junior synonym of *T. annectens*, not only because its morphology is indistinguishable from the latter and distinct from other *Tropheus* species, but also because of the increasingly accepted theory that similar forms on opposite shores of the lake are in fact populations

of a single species that inhabited a smaller paleo-lake in earlier days (Konings, 1992; Konings & Dieckhoff, 1992).

Tropheus duboisi

T. duboisi is probably the most popular species of its genus, most likely due to the coloration of juveniles: pitch black with light-blue to white spots. In comparison to other *Tropheus* species, *T. duboisi*'s mouth is more terminal, and thus its body forms a greater angle with the substrate when feeding than that of *T. moorii*. The latter browses in an almost horizontal position, which probably gives it more stability in the sometimes turbulent upper regions of the rocky coast. This could be a reason why *T. duboisi* is often found at somewhat deeper levels than the *T. brichardi* or *T*. sp. "black" with which it shares the habitat. Nevertheless *T. duboisi* prefers the upper rocky habitat where it feeds on the algae attached to the rocks.

The patchy distribution of *T. duboisi* may be another indication that it represents a less-specialized species. *T. duboisi* occurs at several localities in the northern half of the lake but is nowhere found as the single representative of the genus. In the Congo it occurs near Pemba (Bemba), where it shares the habitat with the "Orange Moorii" (*T*. sp. "black"). It has its largest continuous range in Tanzania where it occurs between the Malagarasi River delta and the border between Burundi and Tanzania. South of the Malagarasi River it occurs at Maswa and Cape Kabogo, and an isolated population has been discovered at Halembe (Konings, 1998). The individuals of these three populations are characterized by a very broad yellow bar on the body and are generally called "Maswa Duboisi" (Maswa is the name of an old fortress, later monastery, located on the shore south of the Malagarasi delta). A different population inhabits the rocky habitat at Karilani Island. This form has a narrow white bar on a black body. The juveniles of all populations look identical and there is no doubt that all known populations belong to a single species.

Kohda & Yanagisawa (1992) found that *T. duboisi* is better adapted to feed from sedimented areas than is *T*. sp. "black" (the authors use the name *T. moorii* for the black *Tropheus* in the north). In areas where both species are found sympatrically, at Pemba in the Congo, it appears that

T. duboisi occupies the deeper and more sediment-rich regions of the rocky biotope. *T.* sp. "black" was found in the extreme shallows, to a maximum depth of 18 metres but most abundant in the upper two metres, while *T. duboisi* was most abundant between 6 and 10 metres but occurs to a depth of about 30 metres. *T.* sp. "black" at Luhanga, 12 km north of Pemba, is found in deeper regions (*T. duboisi* does not occur there) but here they are smaller than conspecifics in shallower areas. *T. duboisi* is larger than *T.* sp. "black" in the deeper, sediment-rich zones because it is able to feed better in such areas, and therefore it has a better chance establishing territories in deeper areas. In shallower areas, where both species overlap, *T.* sp. "black" finds better feeding grounds and grows larger than *T. duboisi* found at the same level, which may give it an advantage in defending a territory. A very similar scenario has been discussed regarding *T. moorii* and *T. brichardi* south of Kipili (see page 55). Along the Tanzanian shores *T. duboisi* is found sympatrically with *T. brichardi*, and, again, the latter is found mainly in the upper three metres of the rocky habitat. Here too it appears that *T. duboisi* is the loser when it comes to establishing territories in the upper rocky region.

In order to explain the disjunct distribution of *T. duboisi* we must accept that the present populations are relicts from bygone days when *T. duboisi* enjoyed a much wider range. Competition from other herbivores (not necessarily other *Tropheus* species) may have pushed it to less-preferred regions of the rocky habitat in some places, and at others may even have pushed it to complete extinction. DNA investigations also suggest that *T. duboisi* is older than other *Tropheus* species tested (Nishida, 1997).

Distribution patterns

Prompted by the existence of similar populations of the same species on opposite sides of the lake (Konings, 1992), I venture to speculate about a possible evolutionary scheme that could have led to the present-day distribution of *Tropheus* species.

T. duboisi is probably the oldest species of the genus. This species is nowadays restricted to a few disjunct locations in the northern half of the lake, but it is likely that it had a much wider distribution in earlier days. A subsequent decline in its distribution could have been caused by two obvious factors: 1. a reduction in the preferred habitat and/or 2. competition from species with similar requirements. Although its habitat preference has not been thoroughly investigated, insofar as it is possible to establish a real preference and not just a tolerance of a certain habitat, we may assume that *T. duboisi* prefers the upper part of the rocky habitat where it finds food in relative abundance. Probably because of the high competition in shallow water it stays in somewhat deeper regions, at the margin of the "preferred" habitat.

The upper regions of most rocky areas in the lake are "ruled" by species of the genus *Petrochromis*. These are relatively large algae-feeding cichlids which are very aggressive in defending feeding areas. Although their aggression is concentrated against conspecifics and other *Petrochromis* species, it is also directed towards most other algae-feeding cichlids. *Tropheus* species, although often described in aquaristic literature as the most advanced or most abundant cichlids of the rocky habitat, play a rather subordinate role

---

page 68   The distribution of *Tropheus annectens* encompasses both sides of the lake.
Tropheus sp. "mpimbwe" is found near Cape Mpimbwe on the central east coast of the lake.
Map: red = range of *T. annectens*; yellow = range of *T.* sp. "mpimbwe".

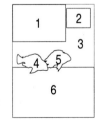

1. *Tropheus duboisi* (Maswa variant) at Halembe, Tanzania.
2. A juvenile *T. duboisi* at Karilani Island (Tanzania); in this individual the vertical bar of the adult becomes superimposed on the polka-dot juvenile pattern.
3. Three adult *T. duboisi* have a "run-in" at Karilani Island.
4. and 5. The polka-dot pattern of juvenile *T. duboisi* has made this species one of the most coveted cichlids of all time.
6. The emperor of the *Tropheus*: *Tropheus* sp. "ikola", also known as the "Kaiser Moorii".

page 69

② *T. annectens*, Kekese

① *T. annectens*, Karilani Is.

② *T. annectens*, Kekese

③ *T. annectens*, M'toto

④ *T. sp.* "mpimbwe", Kalila

③ *T. annectens*, M'toto

⑤ *T. sp.* "mpimbwe", Msalaba

⑥ *T. sp.* "mpimbwe", Korongwe

when it comes to "sharing" the habitat. And it may well be that *T. duboisi* is not strong enough to resist (or tolerate) the *Petrochromis* pressure.

For years it was taken for granted that the dark-coloured *Tropheus* found in the northern half of the lake was a form of *T. moorii* (Brichard, 1989; Kuwamura, 1986a, 1987b; Takamura, 1984; Yamaoka, 1997). This is very unlikely because at Lyamembe near Sibwesa, Tanzania (see map page 13), the latter is found sympatrically with the black *Tropheus*. The rather disjunct distribution of *T.* sp. "black" can be explained in several ways, but the most obvious (at least in my opinion) is that, like *T. duboisi*, the species had a much wider distribution in earlier days. The term "earlier days" is a generalisation; the actual time, and the state of the lake at that time, are both irrelevant to the argument. During the course of time the black *Tropheus* has probably been out-competed at several localities by more aggressive or better adapted cichlids.

The (disjunct) distribution of *T. moorii* is restricted to the southern half of the lake. The population at Sibwesa is probably the remainder of a once more extensive distribution. The forms found from Moba to Kapampa in the Congo are variants of *T. moorii*. In particular the populations north of Kapampa resemble that found near Sibwesa.

T. brichardi differs in several respects from the other members of the genus, the most noticeable difference being the juvenile colour pattern. Juvenile *T. brichardi* exhibit a pattern of vertical bars, each of which is distinctly delineated on the light background. Juveniles of most other *Tropheus* species have bars as well, but these are only partly visible or are so wide that they do not look like bars, the light background colour appearing only as a thin vertical stripe, barely wider than half a scale. The colour pattern in juvenile *T. annectens* resembles that of *T. brichardi* but the two species can be distinguished by the number of spines in the anal fin: four in *T. annectens* and six (rarely five or seven) in *T. brichardi*. In many geographical variants of *T. brichardi* the barred pattern remains visible in adult individuals, particularly in females.

The second difference between *T. brichardi* and most other *Tropheus* species is its habitat preference (or tolerance): *T. brichardi* also thrives in intermediate habitats, even where the rocks are covered with sediment. *T. brichardi*'s success in popu-

lating the lake is obvious: it is the only species of the genus which is found in the northern as well as in the southern half of the lake, and on both sides.

The limited distribution of *T. annectens* may have been caused by two different factors: 1. it is a relatively young species which has not had time to expand into a larger area, or 2. its habitat tolerance is so limited that it cannot compete with other species in less suitable habitats. The latter possibility is rather unlikely because large schools of *T. annectens* have been observed swimming in the open water of the rocky habitat. At M'toto, Congo, one such school was seen that numbered several hundreds of individuals which could not be chased away from preferred feeding sites by *Petrochromis*. The possibility remains, of course, that other areas bordering its present habitat do not satisfy some unknown specific requirement, but it seems more plausible to me that *T. annectens* is a still young species specialised to live in sediment-free rocky habitats. Such habitats are present at other localities along the lake's shoreline but there has not yet been time for it to expand into these areas.

T. sp. "red", *T.* sp. "ikola", and *T.* sp. "mpimbwe" are relatively young species which have developed at a stage later than that suggested for *T. annectens*. All three species are found in areas which were dry land at the most recent low lake level and thus their evolution must have occurred in conjunction with the rise in water level. None of the three species shows a disjunct distribution. Given the observation that the distributions of many species (and geographical races of species) show a mirror image on opposite sides of the lake (Konings, 1992; Konings & Dieckhoff, 1992), in particular where there are steep rocky coasts, one may well argue that, if there are different species/ variants found at opposite rocky sites, it is likely that this distribution pattern developed after the water rose to its present level. Although an "uneven" distribution may be caused by several factors (such as competition, drastic change of habitat, or migration), in the case of the southern Tanzanian shores it seems likely that *T. brichardi*, apparently the most successful *Tropheus* to date, migrated south and occupied areas which may have been less sediment-laden than they are today and which may earlier have been inhabited by *T. moorii*. A similar process may have taken place in the northern half of the lake, where *T. brichardi* may have taken over habitat previously "owned" by the "Black Moorii", just because it became more sediment-rich or in any other way more acceptable to *T. brichardi* than to *T.* sp. "black". This hypothetical scenario is based mainly on the fact that the water level has changed drastically several times during the lake's existence.

Tropheus in the aquarium

In the aquarium males of all *Tropheus* species are clearly territorial and relentlessly chase conspecific males from their domain. Females are courted continuously and often aggressively chased when they are not ready to spawn. The only correct way to keep these species in captivity is in a group of ten or more individuals. The sexes of *Tropheus* can be relatively easily determined by examining the vents. *T. duboisi* can be kept with any of the other *Tropheus* species but is better kept by itself. Never put different geographical races of *Tropheus* or individuals from different localities together in the same tank as this frequently leads to unwanted hybridisation. *T. annectens* and *T. brichardi* are best kept with just one dominant male in the aquarium. A small male may be added as a replacement whenever the leading male "gives up" his territory. The other *Tropheus* species can be housed with as many males as females.

Feeding these species is the most difficult part of keeping and breeding them. The safest food one can give these vegetarians is Spirulina flake food. If a suitable good quality flake is chosen — absolutely no products of warm-blooded animals among the ingredients — this will be sufficient to get these cichlids into breeding condition. A few other foods are also recommended: *Cyclops*, *Mysis*, and other planktonic crustaceans. It is important that the (frozen) food must feel rough to the touch and not soft or slimy. Feeding red or white mosquito larvae, *Artemia*, beef heart and other soft, easily digestible food can be lethal for these fishes. Do not feed more than once a day. Almost all commercially available dry foods contain lots of protein and fat, and in the aquarium these cichlids can eat more of these nutrients in two minutes than they could eat in an entire day in their natural environment. Algal growth on the aquarium decor, promoted by strong lighting, will provide an additional source of food.

Breeding *Tropheus* species is rather difficult for several reasons. The most important is that the male wants to spawn every minute of the day. Another reason is that females are not easily conditioned. They don't need a lot of food but just the right type. Mouthbrooding females can be removed and placed by themselves in a nursery tank. They can be stripped about 25 days after spawning, but there are indications that stripped fry, because of the lack of a proper imprinting period, become bad mouthbrooding females

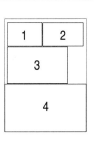

page 72

1. and 2. The so-called "Golden Kalambo" is a xanthic morph of *Tropheus moorii* and occurs only at Chisanza and Isanga Bay in the southeastern part of the lake. Only fully grown adults, males as well as females, exhibit the irregular but colourful pattern. The male in photo 2 is still in the initial phase of the colour change.
The so-called "Red Rainbow" from Kambwimba, Tanzania (3), and the "Sunset Moorii" from Kasakalawe, Zambia (4), are two of the most popular variants of *Tropheus moorii*.

page 73

1. A foraging school of *Tropheus moorii* at Mtosi, Tanzania.
2. *Tropheus* sp. "red" at Cape Kachese, Zambia.
3. The individual on the right is adopting a submissive posture in order to diminish the aggression of the (stronger) individual on the left. *Tropheus* sp. "black" (the so-called "Kirschfleck Moorii") at Mabilibili, Tanzania.
The females in photos 4, 5, and 6 resemble each other and may indicate a close relationship between *Tropheus annectens* (4) and *T.* sp. "mpimbwe" (5: Kalila; 6: Msalaba). Compare these females with the one in photo 7.
7. A female *Tropheus brichardi* at Kapemba, Tanzania.
8. *Tropheus moorii* at Lyamembe is at least 150 km distant from the nearest other population of this species.

when mature. Such females may need up to six spawnings before they hold the eggs to term, while naturally released females usually hold the eggs from the first spawning on. It is therefore better to put mouthbrooding females in nursery tanks and let them take care of the fry until a week after their release. Females may hold up to 20 eggs per spawn.

Another very important point when keeping and breeding these cichlids is never to add new individuals to an existing group. If a breeding group has to be expanded then all the individuals of the new, expanded group must be placed simultaneously in a new aquarium, or one where the decor has been rearranged so as to obliterate previously existing territories.

The *Neolamprologus brichardi* complex

Almost every rocky area along the lake's shores is inhabited by one or more representatives of a group of species which resemble *Neolamprologus brichardi*. This group or complex consists of at least eight different species: *N. brichardi*, *N. savoryi*, *N. pulcher*, *N. gracilis*, *N. falcicula*, *N. splendens*, *N. marunguensis*, and *N. crassus*. Interestingly several of these species can be found sympatrically, and *N. savoryi* is found at every location where any other species of this complex is present. These species seem to be very successful as at most places they are found in large numbers. Most species of this group, which will be discussed in more detail later, prefer depths ranging between 4 and 20 metres. Very rarely are they found in shallower water, and the majority of them occupy a rocky area situated between 7 and 15 metres.

The diet of these elegant cichlids consists of tiny invertebrates found on the biocover, or, most often, in the water column above the substrate. It is not often that one finds small substrate-brooding cichlids feeding from plankton in the open water. The dense populations are sustained by vast amounts of zooplankton, such as *Cyclops*-like crustaceans. The intestinal tract of these cichlids is about 60% of the fish's total length, confirming a carnivorous diet.

Superficially *N. brichardi* and *N. pulcher* seem to behave as schooling fish. Even though they are gregarious they are substrate brooders, and both parents are involved in raising the fry. Sometimes a single male partners two or more females (Gashagaza, 1991). What seems to be a school is in

fact a massive congregation of breeding pairs, sexually inactive individuals, and many juveniles in several stages of development.

N. brichardi and *N. pulcher* have developed an interesting behaviour regarding colony formation and breeding. It appears that the members of a feeding school swim randomly among the group, but when they are disturbed each individual disappears into its own little hole. Each breeding pair has its own territory which is defended by a joint effort. The co-operation of male and female reinforces the pair-bond, and most pairs remain together for much longer than a single breeding episode. Sexually inactive fishes, however, hover further away from the substrate (as they don't feel the urge to stay close to offspring near a breeding cave) than breeding adults, and form a kind of protective umbrella over the breeding section of the community. Approaching danger is first signalled by these scouts and their sudden retreat is a warning signal for other members of the school. This protects the breeding individuals from sudden attacks. The first individuals to be devoured by predators are these sexually inactive "radars", and this will not directly harm the proliferation of the group very much.

The heaviest predation, however, is on juveniles and fry by ambush predators approaching from "behind". Larger lamprologines wander along the nooks and crannies of the rocky habitat and surprise smaller cichlids. Interestingly, the sub-adult and juvenile *N. brichardi* and *N. pulcher* help in the defence of the even smaller fry and larvae. In fact they defend their home site instead of their younger brothers and sisters. Since there are usually juveniles from several consecutive spawns around the nest there are normally also juveniles large enough to fend off possible danger for their younger kin.

Neolamprologus brichardi and N. pulcher

When Brichard first exported *Neolamprologus brichardi* from Lake Tanganyika in 1971 (Brichard, 1989) he gave them the apt trade name of "Princess of Burundi". These elegant cichlids were collected in Burundi near Magara, and although they didn't possess the bright coloration of other, at that time common, aquarium fishes, their marvellous finnage soon made the "Princess" one of the most popular cichlids among aquarists. It proved to be an excellent aquarium resident, even though it is a territorial species which defends its domain with quite some persistency. Given the right amount of room a pair will soon breed and produce successive broods of offspring. This is the most pleasant part of maintaining this species because, in captivity too, older juveniles actively help their parents in defending younger fry.

In 1952 the "Princess of Burundi" became scientifically known as a subspecies of *N. savoryi*, namely *Lamprologus savoryi elongatus* (Trewavas & Poll, 1952). In 1974 Poll raised the "Princess of Burundi" from subspecies to species level. Realising that the name *L. elongatus* was already occupied — *Lamprologus elongatus* (now *Lepidiolamprologus elongatus*) had been described by Boulenger in 1898 — he named the "Princess of Burundi" *Lamprologus brichardi* in honour of the late Pierre Brichard, who had discovered numerous Tanganyika cichlids. In 1985 Colombé and Allgayer published their revision of the genus *Lamprologus*, in which the "Princess of Burundi" was moved to *Neolamprologus* and Boulenger's *L. elongatus* to the revived genus *Lepidiolamprologus*.

Simultaneously with the description of *L. savoryi elongatus* another subspecies was described, namely *L. s. pulcher*. The difference between the two subspecies is based mainly on the pattern of the markings on the gill-cover. Both species have two dark bars between the eye and the outer edge of the gill-cover. In *N. brichardi* these bars have roughly the shape of the letter T lying on its side (see photographs). In *N. pulcher* the bar directly behind the eye is not horizontal but curves downward and does not merge with the vertical bar on the edge of the gill-cover. N. brichardi-like cichlids have been exported from Zambia, the southern part of the lake. Some of these fishes were sold as *Lamprologus pulcher*. It is, however, not known whether these were caught at or near the type locality (which is unknown, but may have been near Mpulungu) or just look like the holotype of *N. pulcher*.

Over the years I have observed a good number of different populations of *N. brichardi*-like cichlids all around the lake and have been able to photograph them as well. Comparison of such pictures has showed me that those gill-cover markings differ slightly between all populations (see also accompanying photos on pages 76 and 77).

N. pulcher and *N. brichardi* resemble each other in shape and behaviour. In captivity both species appear to consider each other conspecific, even when pairs of one species are kept together with those of the other. Soon hybridization will take place and these hybrids are again fertile. *N. brichardi* and *N. pulcher* have never been found sympatrically and although it may seem that the gill-cover markings are a straightforward key to distinguishing them as different species, there are populations where the markings appear to be intermediate between the two species. The form found at Cape Chaitika in Zambia, for instance, is classified as *N. brichardi*, but on close inspection its markings are intermediate (see photo page 76). Just a few miles south of Cape Chaitika, at Kapemba, there is a population of *N. pulcher*, but the markings of that form are likewise not typical of that species (e.g. it lacks the blue colour between the bars). Both forms are, however, quite distinct, which suggests that their assignment to two different species is justified.

Brichard (1989) described several species of this complex, and the types of these taxa are deposited in the Museum voor Midden Afrika, Tervuren, Belgium. The four species, reportedly all from the Congo (but see later), are *N. splendens*, *N. gracilis*, *N. crassus*, and *N. olivaceous*. In his description of *N. olivaceous* (also spelled olivaceus in the same publication) Brichard (1989) mentions the bay of Luhanga (a name known only from the northern part of the lake so he probably meant Lunangwa, which is near Masanza) as the type locality. At Kiku, which is north of the Lunangwa River, I could find only *N. brichardi* and *N. marunguensis*. The photograph on page 375 of Brichard's book shows a specimen of *N. olivaceous* which agrees with the preserved types and with the *brichardi*-like cichlids I found at Cape Tembwe, Kitumba, and M'toto. It is therefore possible that either the

page 76	Geographical variants of *Neolamprologus brichardi*. Map: red = range of *N. brichardi*; orange = the population found in this area (16) seems to be an intermediate between *N. brichardi* and *N. pulcher*, judging from the gill cover markings.
page 77	The distribution of *Neolamprologus pulcher*. Map: red = range of *N. pulcher*, brown = range of *N. splendens*.

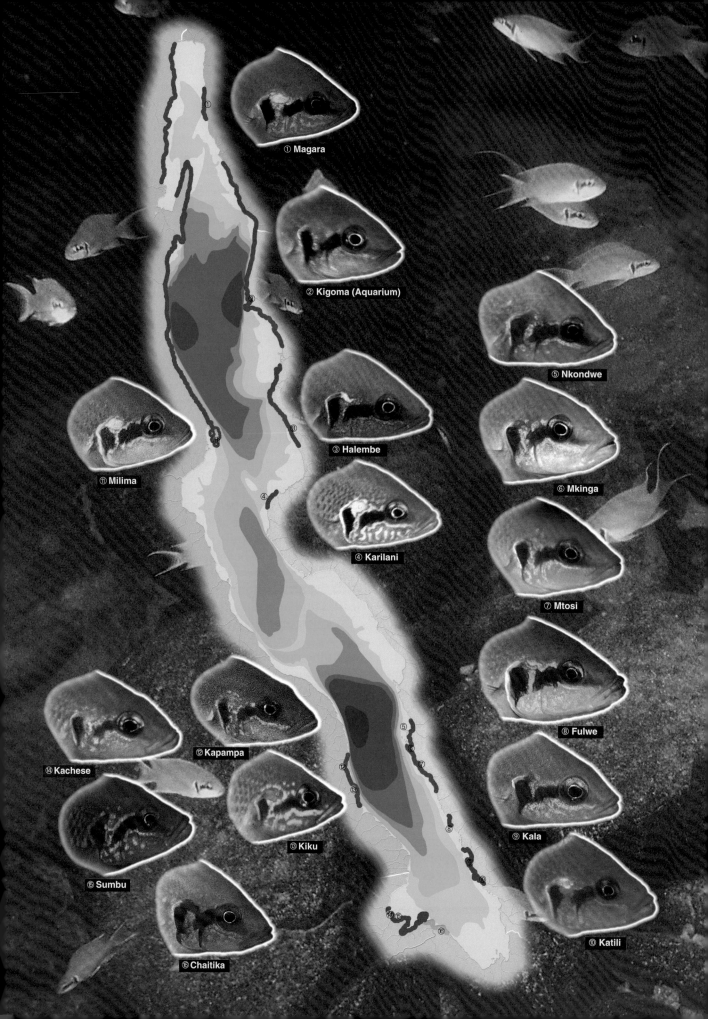

① Magara

② Kigoma (Aquarium)

⑤ Nkondwe

③ Halembe

⑥ Mkinga

⑪ Milima

④ Karilani

⑦ Mtosi

⑧ Fulwe

⑭ Kachese

⑫ Kapampa

⑨ Kala

⑮ Sumbu

⑬ Kiku

⑩ Katili

⑯ Chaitika

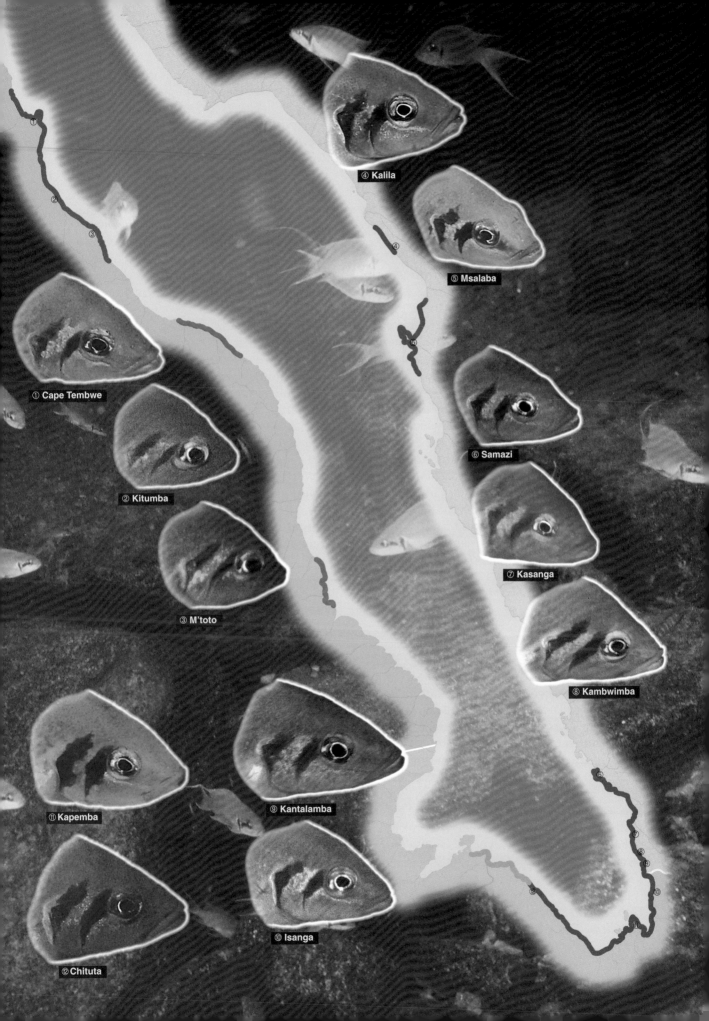

① Cape Tembwe

② Kitumba

③ M'toto

④ Kalila

⑤ Msalaba

⑥ Samazi

⑦ Kasanga

⑧ Kambwimba

⑨ Kantalamba

⑩ Isanga

⑪ Kapemba

⑫ Chituta

type locality lies south of the Lunangwa River (which could not be confirmed by Heinz Büscher (pers. comm.)) or the designation of the type locality as Lunangwa (Luhanga) is an error. Since all his specimens were collected, as mentioned in his descriptions, between Kalemie and Moliro, it is likely that his *N. olivaceous* was collected between Cape Tembwe and M'toto.

The characteristic markings of *N. olivaceous* consist of two vertical curved bars on the gill-cover (like chevrons). The bar directly behind the eye borders the pre-operculum just as the second bar borders the outer edge of the gill-cover (see photos page 77). This pattern is characteristic of *N. pulcher*. Although the type locality of *N. pulcher* is unknown, the gill-cover markings of this cichlid are seen in several populations. I have therefore proposed (Konings, 1993) that *N. olivaceous* should be regarded as a synonym of *N. pulcher*. This has been confirmed by Jos Snoeks (pers. comm.) who is preparing a revision of this complex.

N. pulcher is a small cichlid (maximum total length about 8 cm) which lives in groups in the somewhat deeper rocky habitat. Most individuals are seen below 5 metres. Juveniles at Cape Tembwe, Congo, have orange dorsal fins. The so-called "Daffodil", found near the Kalambo River delta, at Kantalamba, Tanzania, has similar gill-cover markings and yellow unpaired fins in juveniles as well as in adults. Because of the apparent importance of the pattern of the gill-cover markings, I consider the "Daffodil" a geographical form of *N. pulcher*.

Neolamprologus savoryi and N. splendens

One species in the *N. brichardi* complex is easily recognisable and does not show (as far as is known to date) geographical variation, namely *Neolamprologus savoryi*. This small species, with a maximum size of about 8 cm, is identified by the broad vertical bars on its body. It was found at every rocky locality visited. It prefers a depth between 10 and 40 metres, which is somewhat deeper than that for most other *N. brichardi*-like species, which are seen mainly between 5 and 25 metres. *N. brichardi* and *N. pulcher* occur in large groups. *N. savoryi*, however, is usually found in pairs or solitary, and is frequently seen close to groups of the other species of the complex.

Brichard (1989) reported that at Zongwe, Congo, *N. brichardi* shares the habitat with *N. splendens*. The latter species — its trade name is

"New Black Brichardi" — has distinct markings on the gill-cover (see photo page 81) and can thus easily be recognised and distinguished from *N. brichardi*. But I could not find cichlids with *N. brichardi*-type markings on the gill-cover at Zongwe. The three species of the *brichardi* complex I found at Zongwe were *N. savoryi*, *N. splendens*, and *N. gracilis*.

N. splendens is a dusky-coloured cichlid. The gill-cover markings are in the pattern of a V and are not much different to those found in *N. savoryi*. Büscher found populations of *N. splendens* at Kamakonde and Kalo, south of Zongwe, the juveniles of which have vertical bars not unlike those of juvenile *N. savoryi*. Juveniles of *N. splendens* near Zongwe also show such vertical barring (Büscher, 1997). *N. splendens* from Zongwe do not have a yellowish colour — the southern populations have — and for this reason (and some minor anatomical differences) he considers the southern populations to belong to a different species. Büscher recently described this form as *N. helianthus* (Büscher, 1997), but it is here regarded as a geographical variant of *N. splendens*.

N. splendens may be more closely related to *N. savoryi* than to *N. brichardi*. Its maximum size is about 8 cm. Although it is common in the rocky habitat, it was not seen in huge numbers as is sometimes the case with *N. brichardi*.

No gill-cover markings

In the last 20 years many different *N. brichardi*-like cichlids have been exported under various names. We have seen "Lamprologus Kasagera", "Daffodil", "Walteri", "Mbitae", "Black Brichardi", "White Tail", "Palmeri", "Cygnus", and others which have not yet been mentioned in the aquaristic literature. All these variants or species are reportedly collected at different locations.

The first report that, as well as *N. savoryi* and *N. brichardi*, another species may inhabit the same biotope, came from Brichard (1989). After many years of collecting at Magara, Burundi, he found that another *brichardi*-like cichlid lives at levels deeper than 15 metres. This species, named *Neolamprologus falcicula* (Brichard, 1989), lives in pairs or in very small groups, usually not far away from large groups of *N. brichardi*. The most important difference between these two species is the lack of gill-cover markings in *N. falcicula*.

The cichlid with the trade name "Lamprologus Walteri", found in Kigoma Bay and at Cape Kabogo, is a geographical race of *N. falcicula*. At present "Lamprologus Cygnus" from Tanzania (Konings, 1991a), is also regarded as a geographical race of this species. The orange-yellow dorsal and anal fins of the juveniles of the Magara population of *N. falcicula* are somewhat reminiscent of the wonderful juvenile colours of the "Cygnus", which is found near Kipili. *N. falcicula* is never seen in large groups except, perhaps, in the case of the "Walteri" in Kigoma Bay, but here, as elsewhere, they hover never more than 10 cm above the substrate, whereas *N. brichardi* may be seen a metre above the rocks. *N. falcicula* has a wide distribution along the east coast and occurs at various places between Magara, in Burundi, and Samazi, in Tanzania near the Zambian border (see distribution map on page 80).

In his book "*Cichlids and all the other fishes of Lake Tanganyika*", first published in September 1989, Brichard mentions other localities where *N. brichardi*-like cichlids share the habitat with a similar species. These localities are on the southwestern shores of the lake. In his book Brichard describes an additional four new species, all from this area. Büscher, who has visited this part of Lake Tanganyika many times, also reports (1989) on a third, or possibly fourth, species sharing the habitat with *N. savoryi* and *N. brichardi* (and another *brichardi*-like cichlid). He described one of these species as *N. marunguensis*, which has previously been regarded as a synonym of *N. crassus* (Konings, 1993), but which now appears to be a valid species (Snoeks, pers. comm.). *N. crassus*, one of the four species described several months earlier by Brichard, seems to have the wrong collecting locality data, and although the type series contains some *N. marunguensis*, the holotype is different (Snoeks, pers. comm.). *N. crassus* is the form found south of Moliro at Cape Kipimbi and Katete in Zambia, and which has been known in the hobby as "Black Brichardi".

Brichard uses several morphometric characters to distinguish between his new species, which is of course the normal practice in such descriptions. When we look at these *N. brichardi*-like species we notice that the gill-cover markings differ from species to species (see photos). Personal observations in the natural habitat indicate that the individuals of any one population do not show any apparent variation in the pattern of these markings. Therefore we can take these markings as a diagnostic feature in the identification of these species.

N. gracilis lacks gill-cover markings and is further recognisable by the very long filaments on the unpaired fins. Some sub-adult individuals have caudal fins almost as long as the standard length of the fish. This is the main feature that sets the species apart from *N. marunguensis*, which is much stockier and lacks the filamentous fins. This species, like *N. gracilis*, lacks the markings on the gill-cover. It is a rather small species with a maximum size of about 7 cm. Its type locality is at Lunangwa, but it is also found at Kapampa and in Moliro Bay, Congo. At Katete and Cape Kipimbi in Zambia a very similar species (variant?) occurs in the rocky habitat. Moreover this form has a stocky body shape and in all respects resembles that of *N. marunguensis*. The noticeable difference, however, is the fact that these Zambian fishes have an opercular spot, by virtue of which the two forms can easily be told apart. Since markings on the gill-cover seem to be important in distinguishing the various forms of the *N. brichardi* complex I would prefer to consider this form as being a valid species.

---

page 80   The distribution of some closely related lamprologines: *Neolamprologus falcicula, N. gracilis, N. marunguensis,* and *N. crassus.* Note that *N. gracilis* and *N. marunguensis* are found on both sides of the lake. Map: blue = *N. falcicula;* yellow = *N. gracilis;* red = *N. marunguensis;* orange = *N. crassus.*

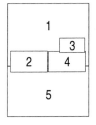

page 81

1. The so-called "Daffodil", *Neolamprologus pulcher*, is found along a very small section of the southern Tanzanian coast at Kantalamba.
2. *Neolamprologus splendens* at Kanoni, Congo.
3. *Neolamprologus savoryi* at Chimba, Zambia.
4. *N. savoryi* at Magara, Burundi.
5. The geographical variant of *Neolamprologus brichardi* at Fulwe Rocks, Tanzania, has a very attractive colour pattern on the cheeks.

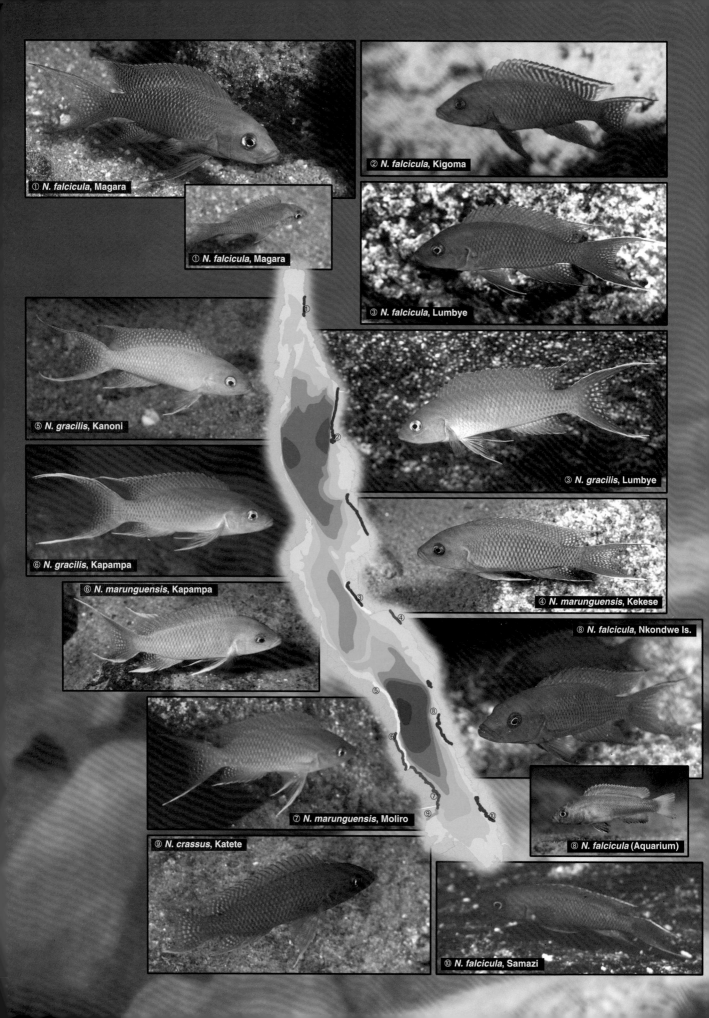

① *N. falcicula*, Magara

① *N. falcicula*, Magara

② *N. falcicula*, Kigoma

③ *N. falcicula*, Lumbye

⑤ *N. gracilis*, Kanoni

③ *N. gracilis*, Lumbye

⑥ *N. gracilis*, Kapampa

④ *N. marunguensis*, Kekese

⑥ *N. marunguensis*, Kapampa

⑧ *N. falcicula*, Nkondwe Is.

⑦ *N. marunguensis*, Moliro

⑧ *N. falcicula* (Aquarium)

⑨ *N. crassus*, Katete

⑩ *N. falcicula*, Samazi

Snoeks (pers. comm.) discovered that the holotype of Brichard's *N. crassus* has a very similar marking on the gill-cover. Since such a pattern (see photo page 80) has not been found in any other species of the *N. brichardi* complex, I suggest that Brichard's type locality data for *N. crassus* — second bay north of Masanza — should be considered erroneous and replaced by Cape Kipimbi. Brichard collected the "Black Brichardi" here in earlier days, and he may have confused specimens from this area with some collected later in the Congo.

Interestingly, on the opposite side of the lake, between Kekese and Isonga (north of Ikola), the rocky habitat is also inhabited by a *marunguensis*-like cichlid. Although *N. gracilis* occurs north of this site, along the Mahali Mountains, I would prefer to assign this form to *N. marunguensis*, because of its stocky shape. When compared to the slender *N. gracilis* at Sibwesa and further north along the central eastern shore, it becomes apparent that the form at Kekese-Isonga is a different species.

At Kapampa *N. marunguensis* and *N. gracilis* are found sympatrically with *N. brichardi* and *N. savoryi*. The first two species lack gill-cover markings, but *N. marunguensis* has broad white-blue edges to the dorsal and anal fins (see photo page 80). The filamentous tips of the caudal fin are white in both species although they are much longer in *N. gracilis*. The latter species is more elongate than the more stockily-built *N. marunguensis*. At Kapampa *N. marunguensis* lives at a rather deep level of the rocky biotope. Büscher (1989) reports that it was found mostly between 25 and 35 metres. *N. gracilis* was also observed at the same location but in shallow water, at about 10 metres.

"Lamprologus White Tail" or "Lamprologus Palmeri" occurs between Sibwesa and Kasoje, Tanzania. This cichlid is without doubt a geographical race of *N. gracilis*. Interestingly, at Lumbye I found it together with another *brichardi*-like species, probably a geographical form of *N. falcicula*. Both these species lack markings on the gill-cover, a situation rather similar to that found across the lake at Kapampa. However, at Lumbye the only difference between the two species appears to be that *N. falcicula* has a darker body than *N. gracilis* and that they are found a little deeper, between 15 and 25 metres. When we compare this form of *N. falcicula* to *N. marunguensis* found further south along the eastern side of the lake, we may notice that it has a more elongate body and a darker colour. However, the members of all other known populations of *N. falcicula* have a thin black edging to the dorsal and anal fins; the form at Lumbye has not. Moreover, juveniles of the Lumbye form were not seen, at least not at a size where they may exhibit some colour in the dorsal fin which could suggest whether we are dealing with a geographical variant of *N. falcicula* or with an undescribed species. For the time being the Lumbye form is treated as *N. falcicula*.

At all other known localities where three species of the complex share the habitat, there are *N. savoryi*, a *N. brichardi*-like cichlid with markings on the gill-cover, and one without any such markings. It thus seems that the possession of these markings or the lack of them is the main feature which segregates the species in the complex (apart from *N. marunguensis* and *N. gracilis* at Kapampa and *N. gracilis* and *N. falcicula* at Lumbye).

Distribution

N. brichardi seems to have the widest distribution of all the species in the complex. I regard all *brichardi*-like cichlids with T-shaped markings on the gill-cover as geographical variants of *N. brichardi*. V-shaped markings are found in *N. splendens*, and chevron-shaped markings in *N. pulcher*. *N. crassus* has a narrow, vertically-elongated, spot on the edge of the gill-cover. *N. falcicula* is recognisable by the lack of markings on the gill-cover and by the dark ground colour. *N. gracilis* lacks the markings as well, but has white edges to the unpaired fins and a light-coloured body. It differs from *N. marunguensis* in its elongated body, the depth of which is 25 to 28% of the standard length (Brichard, 1989). The comparable percentage in *N. marunguensis* is 31 to 35%. It is interesting to note that thus far *N. brichardi*, *N. gracilis*, *N. marunguensis*, and *N. pulcher* have a discontinuous distribution and are found on the west as well as on the east coast of the lake. Populations of *N. falcicula*, or closely related species, are found only along the eastern shores of the lake, and this species also seems to have a discontinuous distribution.

Apart from *N. savoryi*, which occurs at all rocky shores, the species of the *N. brichardi* complex are found in different parts of the lake. In the northern part we find mainly *N. brichardi*, in some places accompanied by *N. falcicula*. In the central part of the lake we find, besides *N. brichardi* and *N. falcicula*, *N. gracilis* and *N. marunguensis* (Tanzania) and *N. pulcher* (Congo). In the southern section of the lake we find all species known in this complex, with a maximum of five sympatric species (*N. savoryi*, *N. brichardi*, *N. marunguensis*, *N. splendens*, and *N. gracilis*) at Kiku, Congo (Büscher, 1997).

Neolamprologus leleupi and its allies

Members of the next group of lamprologine cichlids are usually found in the recesses of the rocky biotope: *Neolamprologus leleupi*, *N. cylindricus*, *N. mustax*, and several forms which are currently assigned to *N. leleupi*. They do not depend on plankton and live a more reclusive life. The members of this group feed on invertebrates, mainly shrimps and other crustaceans, found in the biocover on the rocks or in the cracks between them. A foraging member of this group covers a large area of terrain; it is not known if they have a specific feeding territory; juveniles have also been observed feeding on their own.

Their solitary behaviour may explain their pugnacious attitude towards conspecifics in aquaria. Only ripe females are tolerated in the male's domain. Eggs, however, are deposited in the female's cave. In the lake, therefore, a wandering male may find a ripe female in her cave and spawn with her. In the aquarium these species form a pair during the breeding period. I have not yet seen breeding pairs in the lake but it is possible that the male stays with the female until the young are big enough to face the outside world on their own. In the aquarium the pair bond rarely lasts longer than two weeks.

The holotype of *N. leleupi* was collected in the northern part of the lake at Luhanga in the Congo, and the first individuals exported for the aquarium trade were collected in this area. In the mid seventies Misha Fainzilber exported *N. leleupi* from the Tanzanian east coast of the lake. This geographical variant was later described as *Lamprologus leleupi longior* (Staeck, 1980). Much earlier Matthes described another subspecies and named it *L. leleupi melas* (Matthes, 1959a) because it was dusky coloured and lacked the bright yellow. This subspecies was found at the same locality as *N. leleupi*, namely the northwestern Congolese coast, near Pemba (Bemba). Later on, under-

water observations revealed that such brown-black individuals are also found on the east coast sharing the habitat with the yellow subspecies. In 1986 Poll, in his revision of the Tanganyika cichlids, placed the subspecies *melas* in synonymy with *N. leleupi leleupi* and gave the subspecies *longior* the status of a species.

The fact that a dusky as well as a yellow morph is found within one population (polychromatism) is remarkable, but not unique among cichlids. Poll (1986) found, after examining fresh specimens of the dark morph, that the yellow pigment of such individuals is obscured by the black pigment. The yellow individuals are thus just lacking (temporarily?) the black colour. In Lake Tanganyika polychromatism is known to occur in other species as well. *Ctenochromis benthicola* has orange as well as brown females (Mireille Schreyen, pers. comm.). Büscher (1991b) reports on polychromatism in *N. pectoralis*, and I have found that *N. mustax* also exhibits polychromatism. Cichlids exhibit polychromatism in other water systems as well. In Lake Malawi such individuals are generally described as orange-blotched (OB) or orange (O) and occur in at least 12 species. Among the heroines in Central America orange individuals are known to occur in several species, e.g. *Vieja fenestrata*, *Amphilophus citrinellus*, and *Petenia splendida*.

The yellow *N. leleupi* is thus one morph of a polychromatic species. At several locations it seems that the yellow morph is more frequently

---

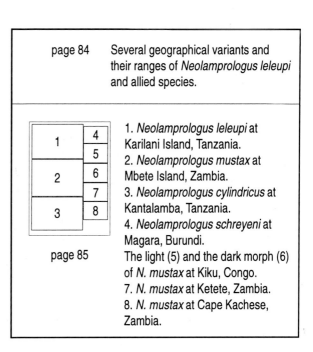

| page 84 | Several geographical variants and their ranges of *Neolamprologus leleupi* and allied species. |

1. *Neolamprologus leleupi* at Karilani Island, Tanzania.
2. *Neolamprologus mustax* at Mbete Island, Zambia.
3. *Neolamprologus cylindricus* at Kantalamba, Tanzania.
4. *Neolamprologus schreyeni* at Magara, Burundi.

page 85   The light (5) and the dark morph (6) of *N. mustax* at Kiku, Congo.
7. *N. mustax* at Ketete, Zambia.
8. *N. mustax* at Cape Kachese, Zambia.

② *N. leleupi*, Maswa

③ *N. leleupi*, Halembe

① *N. leleupi*, Milima

① *N. leleupi*, Milima

④ *N. leleupi*, Bulu Pt.

⑧ *N. leleupi*, Cape Tembwe

⑤ *N. leleupi*, Karilan

⑨ *N. leleupi*, M'toto

⑥ *N. leleupi*, Lumbye

⑩ *N. cf leleupi*, Kapampa

⑪ *N. cf leleupi*, Kiku

⑦ *N. leleupi*, Kekese

N. leleupi
N. cf leleupi
N. mustax
N. schreyeni
N. cylindricus

seen than the dark one, although *N. leleupi* is a rare species at most locations. Kuwamura (1987b), however, reports that *N. leleupi* is common along the central eastern coast near Miyako, Tanzania (about 15 km south of Karilani Island). The dark morph is, of course, less conspicuous than its yellow counterpart, but it does not seem to occur at all localities where yellow individuals are found. At Cape Tembwe, Congo, I found several yellow individuals of *N. leleupi* but no dark ones. The same is true for two localities south of Cape Tembwe, Kitumba and M'toto. This does not mean that the dark morph is not present in these populations; it may simply provide an indication of its low abundance.

N. leleupi appears to have a much wider distribution along the eastern shore. The first specimens collected for the hobby came from Karilani Island, Bulu Point, and the nearby mainland coast south of Karilani. At these locations I also found dark-coloured individuals. The rocky habitats north of Karilani, at Halembe and even as far north as Maswa, are inhabited by *N. leleupi*, although in small numbers. The form at Halembe has a dark body and yellow fins; the one at Maswa has a light-coloured body. I found *N. leleupi* south of Karilani as well: along the Mahali Mountain range and along the range between Isonga and Kekese. The form that occurs here doesn't seem to have yellow-coloured individuals; the few specimens seen (at Kekese and at Lumbye) were either a light silvery-beige or a dark grey-brown colour (see photos page 84).

Some populations of *N. leleupi* (or a very closely related species) along the southern coast of the Congo do not include yellow-pigmented individuals but silvery-beige ones. In these populations one also finds black individuals and light-coloured fish. Around Milima, an island of the Kavala group, I observed a species with a close resemblance to *N. leleupi*. The only obvious morphological difference from known *N. leleupi* appears to be its long pelvic fins. However, in behaviour, polychromatism, and abundance it resembles *N. leleupi*. Again, a very similar population occurs at Kapampa and Kiku. This population also lacks the yellow morph, but black specimens are found sympatrically with light beige individuals. These cichlids seem to have a deeper body than *N. leleupi* from Pemba in the north of the lake or those from Karilani Island and Bulu Point along the Tanzanian shores. *N. mustax* is

closely related to *N. leleupi* and has a deeper body; however, at Kiku it was found sympatrically with the *leleupi*-like species.

In my personal opinion *N. leleupi* is quite a variable species with a rather broad distribution in Lake Tanganyika. The populations with the yellow colour morph are probably old because they are found on both sides of the lake. They were probably present in the paleo-lakes when the water level was much lower than at present. With the rising water level the main population became split up but remained on the west and east central coasts. The species which has been described as *N. longior* is, in my view, a population of *N. leleupi*.

The Malagarasi river delta was, and still is, a barrier to the northward expansion of the species on the east coast. *N. leleupi* is found only south of the river. Its equivalent north of the Malagarasi delta may be found in *N. schreyeni*, which I have seen only at Magara but which may have a wider distribution. *N. schreyeni* looks like a black morph of *N. leleupi* but it is much more reclusive than its conspicuous yellow cousin. It is usually found at levels deeper than 20 metres.

South of Karema the equivalent of *N. leleupi* is *N. cylindricus*, which is very closely related to *N. leleupi* and is found from Karema to Isanga in Zambia. It is, however, not found at Chituta; here it is again "replaced" by *N. mustax*. *N. cylindricus* appears to prefer the shallower rocky habitats and is normally found in the upper 10 metres of this biotope. *N. leleupi* occurs, at some places, at much deeper levels: e.g. at Cape Tembwe or at Bulu Point it is usually found below 20 metres of depth. *N. mustax* also prefers the deeper parts of the rocky habitat. At islands, however, the occurrence of both *N. leleupi* (Karilani Island) and *N. mustax* (Sumbu Island) seems to have a much wider depth range.

N. cylindricus does not exhibit polychromatism, but shortly after the species was discovered and exported for the hobby a so-called "Golden Cylindricus" appeared on the market. Adults of this morph have a yellowish colour between the dark brown bars but juveniles are clearly different from the "regular" morph: a bright yellow patch on the nape, extending onto the body, adorns juveniles. Unfortunately this colourful patch disappears when the fish matures. It is currently not known where this form was collected but the population with the yellowest adults is found at Isanga, Zambia, at the southernmost point of the species' range.

There is a also a close relationship between *N. leleupi* and *N. mustax*. *N. leleupi* at M'toto (see photo) has a relatively deep body and, remarkably, a white chin, a feature that was thought to be typical of *N. mustax*. The northernmost distribution of *N. mustax* is at Kiku, which is about 100 km south of M'toto. *N. mustax* also exhibits polychromatism, and at most locations one can find grey-brown and yellow to orange coloured individuals. In the aquarium *N. mustax* appears to be the most difficult species in this group to keep and to breed. It is not known whether *N. mustax* has larger feeding territories than the other species. Population densities at various places seem to be similar to those found in *N. leleupi*. *N. cylindricus* is the most common species of this group, but this species too is nowhere abundant.

Compressed cichlids

Altolamprologus compressiceps, A. calvus, and *A. fasciatus* are highly specialized cichlids (for discussion of the name *Altolamprologus* see page 22). The first two species have a deep and laterally very compressed body which permits them to enter narrow cracks and shallow caves (after turning the body through 90 degrees about its horizontal axis). *A. compressiceps* and *A. calvus* feed mainly on shrimps and other crustaceans. I have seen *A. calvus* chasing juvenile cichlids as well. *A. fasciatus* has a much more elongate body and feeds predominantly on juvenile cichlids. All three species are stalkers, which cruise through the rocky habitat maintaining a distance of between 30 and 100 cm between themselves and the substrate. *A. fasciatus* maintains the largest distance and is sometimes found cruising about a metre above the substrate. It also moves faster than the other two species — its prey moves faster as well! All three species have a typical hunting posture once they have located a prey item. They stop cruising, then tilt their bodies and assume an almost vertical, head down, position, ready to strike. *A. compressiceps* and *A. calvus* strike from a short distance, usually no more than 10 cm, but *A. fasciatus* may quickly snatch prey from as far away as one metre. Sometimes it uses the side of a large rock for concealment, and "slides" down along the rock surface to the cave at the bottom, where it hopes to snatch some juvenile cichlids. This behaviour is also seen in the much bigger Malawi cichlid *Tyrannochromis macrostoma*.

A noteworthy feature of *A. calvus* and *A. compressiceps* is their scales, which are very thick and strong. Thanks to their peculiar shape these fish can enter extremely narrow slits in the rocks and a miscalculation could lead to ending up jammed inside the crack. Fish can usually wriggle themselves loose but not without losing some scales from the flanks. The hard thick scales of these two predators may be an adaptation for such situations. Another possible benefit of such scales is that these two species can easily withstand aggressive attacks by similar-sized fishes by presenting their "reinforced" flanks to the opponent, by bending the body away from the latter and keeping the head high. It doesn't seem to hurt them when an angry fish bites at their flanks. Such behaviour is frequently observed in the aquarium where, due to the restricted space, fishes regularly encroach on someone else's territory. Most substrate brooders recognize *A. calvus* and *A. compressiceps* as a predator and react vigorously upon encounter. In the aquarium they can even become skilled egg robbers; even stealing from mouthbrooders which they would hardly ever encounter in the wild. They wait patiently for the female mouthbrooder to deposit some eggs and, before the female is able to pick them up, snatch the nutritious eggs with a quick dash.

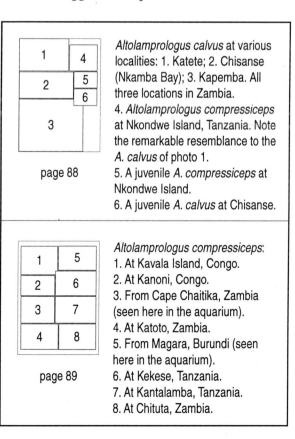

Altolamprologus calvus at various localities: 1. Katete; 2. Chisanse (Nkamba Bay); 3. Kapemba. All three locations in Zambia.
4. *Altolamprologus compressiceps* at Nkondwe Island, Tanzania. Note the remarkable resemblance to the *A. calvus* of photo 1.
5. A juvenile *A. compressiceps* at Nkondwe Island.
6. A juvenile *A. calvus* at Chisanse.

page 88

Altolamprologus compressiceps:
1. At Kavala Island, Congo.
2. At Kanoni, Congo.
3. From Cape Chaitika, Zambia (seen here in the aquarium).
4. At Katoto, Zambia.
5. From Magara, Burundi (seen here in the aquarium).
6. At Kekese, Tanzania.
7. At Kantalamba, Tanzania.
8. At Chituta, Zambia.

page 89

Males of these three specialized predators attain a much larger size — about 15 cm — than females, which grow to a total length of about 9 cm. A similar difference between the sexes is found in other species in which it is mainly the female that cares for the offspring and in which males frequently have a harem. None of the three species is territorial except during the breeding period. The female invariably chooses a hole which is just large enough for her to fit inside, but too small for the male to enter. Very often such holes are located in the thick mineral crust found on almost every object in the lake (see photo page 92), but tiny cracks in the rock and even empty snail shells are also used as brood receptacles. In the area frequented by a male, his domain, one can find several brooding females. When females are ready to spawn they signal this to the male and, even though he cannot enter the female's spawning hole, he is able to fertilize the eggs that are being laid by the female inside. The male discharges his milt over the entrance to the hole, and by fanning with fins and by movements of the female the seminal fluid comes into contact with the eggs. After spawning is completed the female protects the eggs by plugging the entrance of the hole with her body. The male guards the immediate environs for a couple of days but will usually have already abandoned the female by the time the fry appear out of the hole, after an incubation period of about 10 days. The fry are not directly cared for — only the breeding hole is guarded against intruders, including the male. A few days after the fry appear for the first time they wander away from the nest. They have a cryptic coloration and move only in small swimming bouts, thus avoiding attracting attention to themselves. In the aquarium fry of these three species grow exceptionally slowly and it can take three years before they are sexually mature.

Geographical variation

Because of their specialized feeding methods these species are not found in abundance but can be encountered regularly, in particular *A. fasciatus*. There are currently no geographical variants known of *A. fasciatus* even though it is found only in rocky habitats. *A. compressiceps* and *A. fasciatus* have a lake-wide distribution but *A. calvus* is found only in the southwestern part of the lake, between Tembwe (Congo) and Kapemba (Zambia).

There are three different variants of *A. calvus* known. The holotype was collected at Cape Kipimbi on the border between Zambia and the Congo. The form found there belongs to the so-called "Black Calvus", also known as the "Pearly Compressiceps". It has a jet-black body and many silvery-white spots. The black form occurs between Tembwe and Cape Kachese. In Nkamba Bay we find the yellow form of *A. calvus*, and the rocky biotope at Cape Chaitika and Kapemba is inhabited by the white form. Throughout its range *A. calvus* is found sympatrically with *A. compressiceps* and *A. fasciatus*. The anatomical differences between *A. calvus* and *A. compressiceps* consist of the deeper body and shorter snout of the latter.

A geographical variant of *A. compressiceps* found along the east coast between Kipili and Cape Mpimbwe (Msalaba) resembles the black form of *A. calvus* in almost every respect (see photos page 88). It has, however, a deeper body and is thus a form of *A. compressiceps*. Although several geographical variants are distinguished, *A. compressiceps* usually appears very dark in the wild. Even the yellow-orange form at Cape Chaitika and the "Red Compressiceps" from the northern Tanzanian coast appear dark when seen underwater. Only at very short range do they reveal their specific colours.

The first *compressiceps* shipped out of Africa were from Burundi and were later called "Redfin Compressiceps". Red-coloured fins and yellow-coloured bodies are commonly found in the northern and southern populations, whereas the central populations appear to contain mainly black-coloured individuals, sometimes with pearly spots. One of the most outstanding forms is found near the border between Zambia and Tanzania: the so-called "Goldhead Compressiceps". Some (but not all) specimens have a dark body and golden blotches on the head. This form occurs between the Zambian border and Kala in Tanzania. In Chituta Bay and around the islands near Mpulungu a golden *compressiceps* is further characterized by white spots on the body, not unlike those found in *A. calvus*. Brichard (1989) reports another variant along the Congolese shores: the "Orange Compressiceps". Unfortunately this form has never been exported in sufficient numbers to become established in the hobby.

As discussed earlier, females of these three species choose small holes as nesting sites and may sometimes use an empty snail shell. Dieckhoff (pers. comm.) discovered a colony of shell-brooding *A. compressiceps* in Sumbu Bay (Zambia), and this form has subsequently been called "Compressiceps Shell" or *Altolamprologus* sp. "sumbu". The difference between the shell-brooding form and the regular *A. compressiceps* is the fact that males too are small enough to fit inside empty snail shells. Gashagaza *et al.* (1995) found that at least three species of rock-dwelling cichlids have adopted empty snail shells as their home, and also argue that these dwarf forms are not different species but miniature forms of the larger counterparts found in the nearby rocky habitat. *A. compressiceps* is one of the three, while the other two are: «*Lamprologus*» *callipterus* and *Neolamprologus mondabu*. A small shell-brooding *A. compressiceps* has apparently been found near Cape Mpimbwe as well. The local variant of *A. compressiceps* is black and resembles *A. calvus*, and this may have confused the discoverers of the shell-brooding miniature form because it is known in the hobby as "Shell Calvus" (Max Bjørneskov, pers. comm.).

Variabilichromis moorii

The generic name *Variabilichromis* for this species was proposed by Colombé & Allgayer (1985) but later rejected by Poll (1986), and various authors thereafter. Recently Stiassny (1997) published an already classic revision of the lamprologines, based on morphological characters, and came to the conclusion that *V. moorii* should not be grouped together with *Neolamprologus* or *Lamprologus* because of the "primitive" morphology of some of the bones in the skull (basal lamprologine condition of the infra-orbital series). She suggested resurrecting the name *Variabilichromis* for *moorii*, a suggestion which is followed herein.

V. moorii is set apart from the rest of the lamprologines by other characters as well. According to Poll (1956) *V. moorii* has two functional ovaries, which is rare among lamprologines, and according to Yamaoka (1997) *V. moorii* is the only lamprologine that feeds on filamentous algae. *V. moorii* is indeed common in very lush algae gardens in the upper parts of the rocky habitat. Another interesting characteristic of *V. moorii* is the fact that it is not uncommon to find that the female is the larger of the pair — it is not known for certain but it seems that females attain a slightly larger total length than males: 10 cm vs. 9 cm for males. Pairs are commonly seen but harems — not uncommon among lamprologines — have not yet been found. Both male and female defend their offspring. The eggs are often deposited on the vertical face of a rock in the territory. Spawns can number as many as 500 eggs. The tiny fry hover above the nest and feed on plankton as well as on micro-organisms in the biocover.

V. moorii is found in the southern half of the lake, between Kalemie, Congo, and Cape Mpimbwe (Msalaba) in Tanzania. Along the shores of Zambia it is among the most common cichlids found in the shallow rocky habitat. Excepting those of the populations in the Congo, juveniles have a dirty yellow to bright orange colour for the first few months of their lives and turn black when they grow to maturity. Those of the Congolese populations are dark from the start. The juveniles of the populations found around Kipili in Tanzania have the deepest yellow-orange

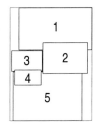

page 92

1. *Altolamprologus fasciatus* at Kantalamba, Tanzania.
2. A female *A. fasciatus* guarding her spawn (inside the hole) at Mkinga, Tanzania.
3. A female *Altolamprologus compressiceps* plugging the small pocket in the mineral crust in which she has laid her eggs (Ulwile Island, Tanzania).
4. A tiny *A. compressiceps* moves circumspectly around the biotope (Magara, Burundi).
5. A female *A. compressiceps* guarding her newly-hatched larvae which can be seen at the bottom of the hole (Kantalamba, Tanzania).

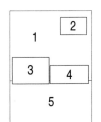

page 93

1. A pair of *Variabilichromis moorii* guarding their fry in Isanga Bay, Zambia.
2. Juvenile *V. moorii* of some populations are a beautiful orange-yellow colour (Ulwile Island, Tanzania).
3. *V. moorii* at Katili, Tanzania. *Neolamprologus furcifer* at Magara, Burundi (4), and at Msalaba, Tanzania (5).

colour. The adult coloration does not seem to differ among the various geographical populations.

Two more lyre-tailed lamprologines

Neolamprologus furcifer can attain 15 cm in total length and lives in the dark recesses of the rocky habitat. It is rather common on the vertical sides of large rocks or on the undersides of overhanging ones. *N. furcifer* is frequently seen hanging upside-down or moving head-down, but its belly is always very close to the substrate. It seems almost as if the fish moves on rails around the rock, paddling only with its large pectoral fins. The female is about a third smaller than the male and is the only parent to take care of the fry. Males have harems and may have more than 10 females "under their control".

Only during spawning is the male allowed to enter the female's territory. After he has fertilized the peculiarly green-coloured eggs the male abandons the female, leaving her with the burden of raising the fry. The tiny fry spread out over the rock surface in the female's territory; they do not hover above the nest as fry of several other lamprologines do — usually those where both parents defend the (more vulnerable) offspring. The fry do not stay with the female for long, and soon disperse to neighbouring rocks.

The diet of *N. furcifer* consists of macro-invertebrates such as crustaceans and insect larvae, which are found in the slits, cracks, holes, and crannies of their habitat. The intestinal tract measures about 50% of the fish's total length (Poll, 1956) and the pharyngeal teeth are slender and sharp, suggesting a predatory appetite. The fact that *N. furcifer* moves in close contact with the substrate may facilitate the location of prey. On the other hand prey would probably consider a free-moving object more dangerous than a mobile "hump" on the rock. *N. furcifer* has very precise control of its swim bladder, probably to permit the precise movements needed for its peculiar locomotion. Experiences in handling this species in captivity suggest that the bladder is more vulnerable to sudden changes in pressure, even if of a minor nature, than that of other lamprologines. Almost all wild-caught *N. furcifer* are initially ill at ease in a tank and suffer problems with their swim bladders. The fishes linger motionless on the bottom and are unable to stay aloft in the water column. A few weeks after their introduction into the aquarium the disfunction seems to have healed in most cases, since the fishes look healthy and are able to stay in any position in the water. This problem may occur even when netting out fishes from one tank to another.

There seems to be very little geographical variation in *N. furcifer* although it is found in almost every rocky habitat of the lake. The individuals of the southern populations have some dark irregular and elongated spots on the body; those of Burundi appear to have a more even-coloured body, but faintly-marked blotches are still visible.

Neolamprologus longicaudatus has been known in the hobby as the "Ubwari Buescheri" because that is where the aquarium specimens hail from (Brichard, 1989). It was described from Cape Banza, the tip of the Ubwari peninsula, but I have found the species around the Kavala islands as well (Konings & Dieckhoff, 1992). It is reasonable to assume that it is also present in the rocky habitats between these two points, which are, nevertheless, about 200 km apart.

N. longicaudatus is characterized by two horizontal lines on a light-coloured body and resembles *N. bifasciatus* in some respects, but the latter does not have a lyre-shaped tail. The trade name may suggest that it is closely related to *N. buescheri*, but it can be distinguished from this species by the fact that is lacks the broad submarginal band in the dorsal fin.

N. longicaudatus can attain a total length of about 15 cm, which is considerably larger than that of *N. buescheri*, and is most often found solitary. It feeds on invertebrates, as most other lamprologines do, and is seen more often outside caves of the rocky substrate than *N. buescheri*, which is only found in the southern region of the lake. The form at the Kavala islands has the lower half of the body and head marked with a dark band, apart from the throat, which is white. The form found around the Ubwari Peninsula is a lighter colour.

Pearly-coloured piscivores

Lepidiolamprologus elongatus

Lepidiolamprologus elongatus, which is very common throughout the lake, is a piscivore which preys mainly on the fry of other cichlids such as *Neolamprologus brichardi* and other substrate brooders. Crustaceans and insect larvae are eaten as well. The intestine measures just 40% of the fish's total length indicating a carnivorous diet. Males can grow to a respectable 20 cm in total length and large adults are rather common.

Although breeding takes place in the protected environment of rocks, the eggs are not completely hidden and the fry are guarded outside the rocks. Juvenile *L. elongatus* consume mainly plankton and hover above the nest. They are guarded for a rather long period by both parents, but a male can sometimes "service" two breeding females at the same time. The fry may measure more than 3 cm before the parents relax their guarding vigour and prepare for a new brood. The young from the previous spawn are tolerated in the nest. This fact is abused by sub-adult *Telmatochromis bifrenatus* that enter the nest, mingle with the larger young, and feed on the eggs (see photos page 97). The egg-robbers are not recognized as such (they look like juvenile *L. elongatus*) and can consume a considerable part of the spawn.

Nevertheless, *L. elongatus* is a very successful species, as is demonstrated not only by their lake-wide range, but also by the fact that large groups, numbering often more than 50 individuals, roam through the habitat. The members of such a group are of approximately the same size — between 10 and 15 cm. Like a pack of hungry wolves they move through the habitat, leaving a trail of devastation behind them. They gobble up everything that is encountered on their way. Such behaviour is even more commonly seen in «*Lamprologus*» *callipterus,* which *L. elongatus* seem to follow (literally).

Another remarkable fact concerning *L. elongatus* is the capture of some specimens at an unbelievable depth of almost 200 m (Poll, 1956). Geographical variation, apart from slight differences in yellow pigment, is not apparent in the populations seen so far; *L. elongatus* appears to be a very homogeneous species.

Lepidiolamprologus kendalli

Lepidiolamprologus kendalli (*L. nkambae* is a junior synonym) resembles *L. elongatus* in shape and coloration, and, like *L. elongatus*, is a piscivore. *L. kendalli* was described from Mutondwe Island (Crocodile Island) offshore of Mpulungu — the holotype of *L. elongatus* was also caught in this area — but has a much smaller range, being found between Nkamba Bay in Zambia and Kala in Tanzania. It is a much rarer fish than *L. elongatus* but

can be found in shallow water as well as deeper than 40 metres. *L. kendalli* is almost always solitary and covers large distances while cruising through the habitat searching for prey. Only once, at Samazi in Tanzania, did I find breeding pairs. There were several pairs all at a depth of about 45 metres. At this level the rocky coast is covered with thick layers of sediment and rocks are visible only as small heaps protruding from the muddy bottom. It appeared that almost every group of rocks had a breeding pair of *L. kendalli*. Females are about 25% smaller than males, which can attain a length of about 22 cm. It is not known whether *L. kendalli* needs subdued light conditions in order to breed or whether the juveniles feed on the micro-organisms found over muddy bottoms. Other reports of breeding *L. kendalli* in the wild are not known to date.

Although a lot of attention has been given to the geographical variants exported for the aquarium hobby, it seems that the so-called "Black Nkambae" and lighter-coloured individuals can be found at the same locality. Geographical variation exists very little, if at all. *L. kendalli* is easily distinguished from *L. elongatus* by the intricate patterning on the head which is lacking in *L. elongatus*.

In captivity *L. kendalli* are difficult to pair up, and not uncommonly the female is chased out of

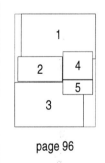

page 96

1. *Neolamprologus longicaudatus* at Milima Island, Congo.
2. *N. longicaudatus* from Ubwari, Congo, in the aquarium.
3. *Lepidiolamprologus elongatus* at Kekese, Tanzania.
4. A plaque of eggs of *L. elongatus* attached to the vertical face of a rock (Kapampa, Congo).
5. A brood-guarding *L. elongatus* at Ulwile Island, Tanzania.

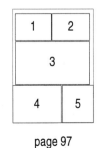

page 97

1. *Lepidiolamprologus kendalli* from Nkamba Bay, Zambia, in the aquarium.
2. *L. kendalli* at Kambwimba, Tanzania.
3. *L. kendalli* at Kala, Tanzania.
4. and 5. Juvenile *Telmatochromis vittatus* feeding on the eggs of *Lepidiolamprologus elongatus*.

the tank — literally! Even very small holes in the aquarium cover are "found" by jumping kendalli. Once a pair has formed regular spawning seems to be the rule. Both *L. elongatus* and *L. kendalli* can produce spawns of more than 500 eggs, but it seems that the clutch size of *L. elongatus* is larger.

Slender *Telmatochromis*

The genus *Telmatochromis* is well-represented in the rocky habitat and most of its members are frequently seen. At every rocky coast there are two species present: *T. temporalis* or *T. sp.* "temporalis tanzania" and either *T. brichardi*, *T. bifrenatus*, or *T. vittatus*. The first two species have a beige-brown coloration and in some individuals (probably those showing a fright colour pattern) a pattern consisting of two horizontal lines is visible. The other three species have a creamy-white body and very distinct dark horizontal lines. *T. brichardi* and *T. vittatus* both have a pronounced midlateral band and one along the base of the dorsal fin. The difference between these two cichlids is the smaller eye and larger adult size of *T. vittatus* (10 cm for *T. brichardi* vs. 6 cm for *T. bifrenatus*). *T. bifrenatus* has an additional black line between the midlateral and dorsal bands. All five species feed occasionally or predominantly on algae. *T. temporalis* and *T. vittatus* seem to feed mainly on filamentous algae, although *T. temporalis* also feeds on killifish eggs. *T. temporalis* bites the algae from the substrate while the other four species "yank" them off by grasping a small bundle of algae filaments, bending their bodies in an S-shape, and, by suddenly straightening their bodies again, shooting forward. The thrust is enough to shear off the algal strands from the rock. This behaviour resembles that of juvenile fish preying on live *Artemia*; they too twist their bodies in order to give the tail the maximum thrust while striking. The difference between *T. temporalis* and the slender algae-pluckers is that the latter have a much narrower mouth and can thus penetrate deeper into smaller pockets in the substrate than can *T. temporalis*. Besides algae the slender species also feed on invertebrates and the eggs of the large substrate brooders such as *Lepidiolamprologus elongatus* and *L. profundicola*. These large lamprologines have enormous spawns which are laid on the vertical face of a rock where they are visible to the outside fish-world. These species are, however, large enough to defend their spawns against

predators — or at least that is what they assume. Since they spawn repetitively at the same location (the female's territory), juveniles of the previous brood are sometimes still around when a new plaque of eggs is deposited. These juveniles are not chased from the nest. Juvenile *T. bifrenatus*, *T. brichardi*, and *T. vittatus* look like young *Lepidiolamprologus* and are thus likewise not chased from the nest. While the large *Lepidiolamprologus* pair chase away all larger intruders, the tiny *Telmatochromis* feed, in full view, on the eggs and are unchallenged by the pair.

All these *Telmatochromis* species are cave brooders which select tiny holes in the substrate as nesting sites. Apart from *T. sp.* "temporalis tanzania", in which the behaviour has not yet been observed, they also use empty snail shells as brood receptacles. In fact *Telmatochromis* seem to be more abundant in and around empty shell beds than in the pure rocky habitat. *T. temporalis* and *T. vittatus* have, in many places, developed into miniature copies of their larger counterparts in the rocky habitat, just so they can fit inside the empty snail shells (see page 216).

The northern form of *T. temporalis* has been formally described as *T. burgeoni* — the holotype hails from Nyanza Lac — and it is not yet clear whether *T. burgeoni* should be regarded as a different species or just as a variant of *T. temporalis*, the holotype of which hails from the southern part of the lake (Tetsumi Takahashi, pers. comm.). Pending a revision of these species *T. burgeoni* is here regarded as a synonym of *T. temporalis*. Likewise the form associated with empty shells is regarded as a dwarf morph of *T. temporalis* (see page 216). The species found along the south-central Tanzanian shore and which I have previously named *T. burgeoni* (Konings, 1996), is different, and is here referred to as *T. sp.* "temporalis tanzania".

T. temporalis has the widest distribution but is not found at every rocky shore. Between Wampembe in the southeastern part of the lake, and the Mahali Mountains in the central part, its place appears to be taken by *T. sp.* "temporalis tanzania". The fish cichlid keepers used to call *T. bifrenatus* has now been described as *T. brichardi*, while the true *T. bifrenatus* has very rarely been exported as an aquarium fish. *T. brichardi* occurs in Burundi near Magara, along the southern half of the Congolese shores, up to Kapampa; along the southern part of the Tanzanian shore, south of

Sibwesa; and into Zambia along the eastern side of Chituta Bay. The remaining area in the south is inhabited by *T. vittatus* (found between Kapampa and Chituta). *T. bifrenatus* occurs at Sibwesa and further north all the way up to the border between Tanzania and Burundi, and is also found along the northern Congolese coast between Kavala and Luhanga.

Thick-lipped cave-dwellers

The members of the genera *Julidochromis* and *Chalinochromis* are closely related. The presence of "warts" on the lips of *Chalinochromis* species is the main characteristic that separates them from *Julidochromis*. Members of both genera use their lips to crush sponges and invertebrates living in the biocover. Kohda & Hori (1993) found that *J. marlieri* feeds for more than 80% of the time on sponges, and Hori (1987) reports that *J. regani* and *Chalinochromis brichardi* feed on sponges as well. Sponges are rather abundant in some places but very few species feed on them. Apart from the species mentioned, *Telmatochromis dhonti* and a *Synodontis* catfish also feed on these rubbery animals which usually grow on rocky surfaces lying in shadow. Because sponges are eaten by only a handful of species, *Julidochromis* and *Chalinochromis* have found a niche in almost every rocky habitat. The tiny warts on the lips of *Chalinochromis* species may be an adaptation towards getting a better hold on smooth-surfaced sponges, although *C. brichardi* does not appear to be specialized in eating sponges, as it eats other invertebrates, and even algae, as well (Hori, 1987).

Apart from having warts on the lips, *Chalinochromis* also differs in another aspect from *Julidochromis*: it has a slightly different breeding technique. *Chalinochromis* and *Julidochromis* species are substrate brooders that hide their spawn in a dark rocky hole. The male guards the small territory around the nest while the female remains inside. When the fry have grown to about a centimetre long they are taken out into the open in *Chalinochromis* species, but stay inside in *Julidochromis* species. The hovering *Chalinochromis* fry feed from the plankton and are protected by both parents. Brichard (1978) found that when young *Chalinochromis* reach double this size they are abandoned by the parents, which probably separate as well. Young *Julidochromis* are never abandoned and are tolerated inside the nest even when a new spawn

is present. They never hover above the nest but stay in close contact with the substrate.

Chalinochromis

Chalinochromis contains two described species, *C. brichardi* and *C. popelini*, and one (perhaps two) undescribed forms. *C. brichardi* was described from specimens collected in Burundi, but the species has a much wider distribution. It occurs in the entire northern half of the lake, north of Kalemie in the Congo and of Halembe in Tanzania. Interestingly, it is also found in the southernmost part of the lake between Cape Kachese and the mouth of the Kalambo River in Zambia. *C. brichardi* seems to be an older species with a lake-wide but disjunct distribution, and with a remarkably constant colour pattern. Some specimens have a dark spot in the soft-rayed part of the dorsal fin, and this occurs in all known populations. It seems that this spot disappears when the fish fully matures, although it has been seen in some large individuals as well.

The rocky habitats in the central parts of the lake are populated with different forms of *Chalinochromis*. The colour patterns of these forms reflect the two stages of the juvenile patterns of *C. brichardi*. Young *C. brichardi* have two horizontal black lines on the flank. During maturation these

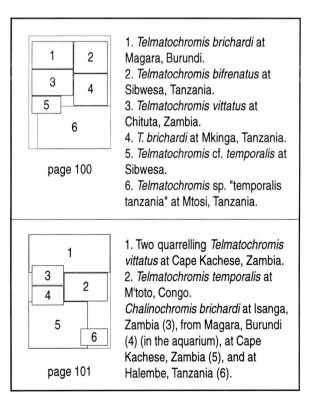

1. *Telmatochromis brichardi* at Magara, Burundi.
2. *Telmatochromis bifrenatus* at Sibwesa, Tanzania.
3. *Telmatochromis vittatus* at Chituta, Zambia.
4. *T. brichardi* at Mkinga, Tanzania.
5. *Telmatochromis* cf. *temporalis* at Sibwesa.
6. *Telmatochromis* sp. "temporalis tanzania" at Mtosi, Tanzania.

page 100

1. Two quarrelling *Telmatochromis vittatus* at Cape Kachese, Zambia.
2. *Telmatochromis temporalis* at M'toto, Congo.
Chalinochromis brichardi at Isanga, Zambia (3), from Magara, Burundi (4) (in the aquarium), at Cape Kachese, Zambia (5), and at Halembe, Tanzania (6).

page 101

lines disappear, but just before the lines completely dissolve they exist as two horizontal rows of dark spots. The latter pattern is "retained" in fully mature *C.* sp. "ndobhoi" whereas the two-line pattern is seen in adult *C. popelini* (also known as "Chalinochromis Bifrenatus").

C. popelini was described by Brichard (1989) from a population found near Moba along the Congolese shore of the lake. The species known in the hobby as "Bifrenatus" hails from the central Tanzanian coast. I believe both forms are conspecific and should be called *C. popelini*. On the western side of the lake this species occurs between Kalemie and Lunangwa, and on the eastern coast it is found between Katili and Kasoje (at the foot of the Mahali mountain range). Within the eastern range there are populations that look different and may belong to a different species altogether. Between Fulwe Rocks at Wampembe and Nkondwe Island near Kipili lives a *Chalinochromis* form which resembles *C. popelini* but lacks the horizontal stripes on the body, has a much darker ground colour, and is also much larger than other *Chalinochromis* variants (see photos page 104). For the time being this is regarded as a variant of *C. popelini* because sub-adult specimens are indistinguishable from that species.

In some Tanzanian populations of *C. popelini* there are individuals with orange or yellow spots on the body. However, such a pattern has not consistently been found in all (adult) members of the population. *Chalinochromis* sp. "ndobhoi", characterized by the dotted lines on the body, occurs on the east coast between Kasoje and Bulu Point and is also found around Karilani Island.

Julidochromis

Three species of *Julidochromis* are found in the rocky habitat: *J. marlieri, J. transcriptus,* and *J. dickfeldi*. The last-named species has a very restricted range in the southwestern corner of the lake, where it is found between Cape Kachese (Zambia) and Moliro (Congo). In habitat preference and feeding behaviour it resembles *C. popelini*, and its colour pattern is also very similar to that of the latter. At Cape Kachese three species of the thick-lipped and slender cichlids group are found: *C. brichardi, J. dickfeldi,* and *J. regani*. Only very rarely does one find more than a single *Julidochromis* or *Chalinochromis* at a specific site. If *J. dickfeldi* were to be regarded as a *Chalinochromis*,

as it seems it may have derived from *C. popelini*, then Cape Kachese would be the only place known where two members of that genus occurred. Two different members of *Julidochromis* at a single location is a very rare occurrence (apart from within the *J. dickfeldi* range), and is known only from the coast at Isanga (Zambia) and Kantalamba (Tanzania) (on either side of the Kalambo River mouth), and at Uvira (Brichard, 1978). At these three sites *J. ornatus* and *J. marlieri* coexist in the same habitat.

J. marlieri has a disjunct distribution in the northern half of the lake and is found at Magara in Burundi, at Luhanga (Poll, 1956), and at the Kavala Islands in the Congo, but may be found along the Congolese shores between the latter two sites as well. *J. marlieri* also has a restricted range in the southern half. Although it occurs at Cape Tembwe, it has not been seen further south along the Congolese shore. In the southeastern part it is more common and is found between Isanga (Zambia) and Kala (Tanzania). Interestingly there are *J. marlieri* populations at Samazi and at Halembe (a beautiful form with orange dots on a black body) which consist of small individuals with a colour pattern resembling that of the species known as *J. transcriptus*. This pattern, however, includes a black bar below the eye, a characteristic of *J. marlieri*, and these populations are therefore regarded as a dwarf form of *J. marlieri*. The same is true of the form at Katoto and that at Kaku, which is better known as "Gombe Transcriptus".

Julidochromis species are very difficult to define, and I believe that in many instances the colour pattern of a particular individual is dependent on the type of habitat it is found in. *J. marlieri* and *J. regani* have an overlapping colour pattern and it is sometimes difficult to assign a specimen to one species or the other except by assuming that vertical barring is restricted to *J. marlieri*. In the sediment-free rocky habitat one usually finds dark-coloured *J. marlieri* or *J. transcriptus*, but where the rocks are covered with sediment you are more likely to find lighter coloured *J. regani* or *J. ornatus*. And if we descend along the rocks to deeper regions we will notice that there is also a vertical gradient in the intensity of the black pigment: in the upper layer light-coloured *Julidochromis* and in the deeper areas dark-coloured ones.

J. transcriptus was described from the population found at the northwestern end of the lake between Luhanga and Pemba (Bemba) (Matthes, 1959b). Brichard (1978) reports that there are light-coloured forms and dark-coloured ones, and also that a yellow coloured variant without vertical barring occurs in adjacent areas, and was exported as *J. ornatus* in the sixties and early seventies (Brichard, 1978). Matthes (1962) re-described *J. ornatus* (the holotype of which originates from the extreme south of the lake, near Mpulungu) using specimens from the extreme north (see page 178 for further discussion). If we compare *J. transcriptus* from the type locality with the small *J. marlieri* forms found in other areas (Halembe, Samazi, Katoto, and Kaku) we notice a difference in the colour pattern on the head. The type does not have black markings under the eye apart from a dark stripe running through the lower half of the eye. Members of the four populations mentioned, which up to now have been referred to as *J. transcriptus*, do have markings on the cheeks and resemble those of *J. marlieri*.

I would like to suggest calling *Julidochromis* forms with vertical bars and with markings on the cheek *J. marlieri*, and those without cheek markings *J. transcriptus*. This would mean that the so-called "Gombe Transcriptus" is a variant of *J. marlieri*, as is a light form found at Katoto. The dwarf forms I found at Halembe and Samazi would have been called "Transcriptus" if they had been available to hobbyists, but are here referred to as *J. marlieri*. The only place, other than the type locality, known to be inhabited by *J. transcriptus* is Kapampa in the Congo. In 1993 I argued that this form should be assigned to *J. ornatus*, but having seen more forms of *Julidochromis* around the lake I must admit that it resembles *J. transcriptus* more than any other species in this genus.

At Kapampa I found a similar situation to that which Brichard (1978) describes for *J. transcriptus* at Luhanga. At Kapampa *J. transcriptus* lives in the intermediate and rocky habitats and has a considerable depth range. Interestingly, individuals with a dark coloration are found together with those having a much lighter colour pattern. The ones living at greater depths are usually much darker than those in the shallows, but this is not a definitive rule: dark specimens were also found in shallow water. A few kilometres on either side of this location I found forms with a colour pattern consisting of three horizontal stripes (see photo page 177) and therefore assigned to *J. ornatus*.

Brichard (1978) reports that *J. transcriptus* in the north can be found in different populations separated by just a few kilometres of sandy beach and alternating with populations of *J. ornatus*. Knowing the great variability *Julidochromis* can show in a single population, and knowing their lake-wide but alternating distribution, one would have to prove that *J. transcriptus* and *J. ornatus* are indeed two different species by finding a locality where both species live sympatrically. Such a locality is not yet known. A similar situation exists for *J. marlieri* and *J. regani*, i.e. there is no location known where both species are found sympatrically.

When young *J. transcriptus* are taken out of a fully decorated aquarium and subsequently placed in a light and bare tank, they change to a very light colour within a few days and can hardly be distinguished from young *J. ornatus*.

The breeding behaviour of the various *Julidochromis* species differs slightly. They are all cave brooders and females usually attach the eggs to the sides or the ceiling of the spawning cave. *J. transcriptus* and *J. ornatus* spawn (with small batch sizes) every fortnight whereas the other species (*J. marlieri*, *J. regani*, and *J. dickfeldi*) breed every four to five weeks. *J. marlieri* and *J. regani* are further distinguished in that females are usually larger in size than males (on an age for age basis). Eggs and fry are looked after by the female in the main. The females of the larger "Julies" fan the eggs more frequently and are in closer contact with their broods than those of *J. transcriptus* and *J. ornatus*.

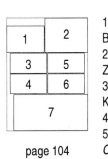

page 104

1. *Chalinochromis* sp. "ndobhoi" at Bulu Point, Tanzania.
2. *Julidochromis dickfeldi* at Katete, Zambia.
3. *Chalinochromis popelini* at Kambwebwe, Congo.
4. *C. popelini* at Kekese, Tanzania.
5, 6, and 7. Three different forms of *C. popelini* at Ulwile Island, Tanzania.

page 105   Several geographical variants of *Julidochromis marlieri* and *J. transcriptus*. The map shows the distribution of various *Julidochromis* species: black = *J. marlieri*; blue = *J. transcriptus*; red = *J. ornatus*; yellow = *J. regani*; purple = *J. dickfeldi*.

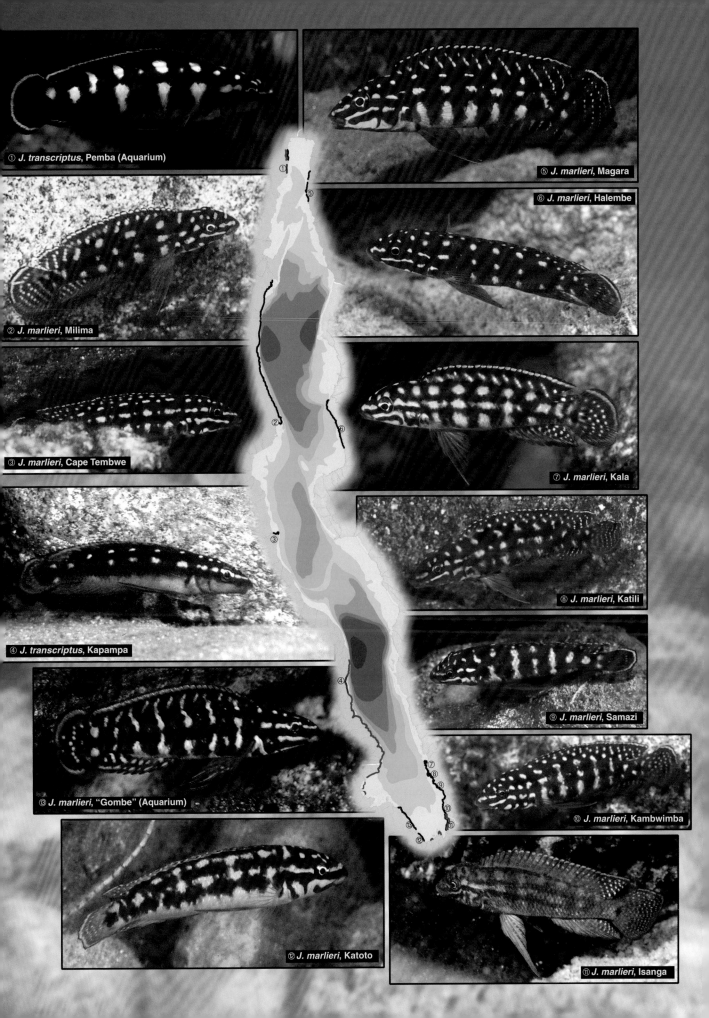

① *J. transcriptus*, Pemba (Aquarium)

⑤ *J. marlieri*, Magara

⑥ *J. marlieri*, Halembe

② *J. marlieri*, Milima

③ *J. marlieri*, Cape Tembwe

⑦ *J. marlieri*, Kala

④ *J. transcriptus*, Kapampa

⑧ *J. marlieri*, Katili

⑨ *J. marlieri*, Samazi

⑬ *J. marlieri*, "Gombe" (Aquarium)

⑩ *J. marlieri*, Kambwimba

⑫ *J. marlieri*, Katoto

⑪ *J. marlieri*, Isanga

Asprotilapia leptura is commonly seen in foraging schools feeding from the biocover of large boulders. The mouth of *A. leptura* is unique among Tanganyika cichlids and is similar to that of *Labeotropheus* from Lake Malawi. *A. leptura*, which grows to a total length of about 10 cm, is usually found with its belly close to the substrate. It feeds by rapidly biting at the Aufwuchs with its underslung mouth. Due to the mouth's position — the mouth opens downward — it can peck at algae from a position parallel to the rock. It does not press its lips against the substrate but rather taps hard on the rock and pulls the algae from the biocover (Yamaoka, 1990). When plankton is abundant *A. leptura* may leave the rocks and form a plankton-feeding group in the water column.

A. leptura is a biparental mouthbrooder. There is no sexual dichromatism, but the male is a bit larger than the female. Before breeding a male and a female separate from the school, and stake out a small territory prior to spawning. In the aquarium the male uses small pits in the sand when available, because he does not construct them himself. During spawning the female releases a few eggs at a time, which are fertilized by the male when they are still on the sand. Different sites may be used during a single spawning sequence. When more eggs are expelled than can be accommodated by the female, the male will assist her by collecting the leftover eggs. Large spawns are mouthbrooded by both parents, but usually it is the female who broods the eggs and early-stage larvae. Halfway through the three-week incubation period the female transfers all or part of the brood to the male. During the incubation of eggs and larvae the pair stay close together, and when the fry are released both male and female guard them. Post-release guarding can last for several weeks, and pairs with relatively large fry are rather common (see photo page 108).

A. leptura has a wide distribution in the lake and is found at almost every rocky coast that contains large boulders. Geographical variation is unknown. Earlier reports that a yellow form occurs in Chituta Bay (Konings, 1988) refer to a different species — *Xenotilapia* sp. "sunflower" (see page 122).

O. boops resembles *O. ventralis* in shape, behaviour, and coloration. The main difference between these two species is that *O. boops* has tricuspid teeth in the outer row whereas these teeth are bicuspid in *O. ventralis*. A small difference one might say, but both species are found sympatrically at Nkondwe Island, Cape Korongwe, and at Cape Mpimbwe. *O. boops* is normally found in the upper parts of the rocky habitat but not in the surge zone, as is *O. ventralis* (see page 39). Territorial male *O. boops* defend a spawning site on top of a large flat boulder, whereas *O. ventralis* defends almost any type of rocky surface as its breeding territory. *O. boops* defends its domain by violently chasing all intruders from a relatively large breeding territory — territories usually measure more than one metre in diameter. *O. boops* has an overall dark-brown to black coloration in the southern part of its range, where it does not share the habitat with *O. ventralis*, but males exhibit bright white to light-blue patches on the caudal peduncle and tail at Nkondwe Island where it shares the habitat with an all-black form of *O. ventralis*. Curiously, at Cape Mpimbwe this situation is reversed: here *O. ventralis* has white patches on a black body colour whereas *O. boops* is completely black again, apart from a few tiny white-blue spots on the tail and dorsal fin. At Nkondwe Island the population of *O. boops* is much larger than that of *O. ventralis*, perhaps because it seems that *O. boops* is more aggressive and extends its territory over a wider depth range than *O. ventralis*. It thus appears that since these two species look so much alike they needed mutually to exclude the possibility of females getting confused when choosing their partners.

O. boops, like *O. ventralis*, feed from the Aufwuchs on the rocks. They don't scrape or comb the algae but non-selectively pick up any loose material from the substrate. Their food consists mainly of algae (diatoms), but sand grains and other indigestible matter are sucked up as well. Females and non-territorial males also feed on plankton when this is available.

O. boops is a mouthbrooder employing a similar breeding technique to that described for *O. ventralis* (page 42). In fact all *Ophthalmotilapia* species breed in a similar fashion and males of all species have yellow egg dummies at the tips of their ventral fins.

O. boops occurs along the eastern shore of the lake between Hinde B, the southernmost point of its range, and the rocky coast south of Karema. The southern populations are characterized by dark brown individuals and the northern by blue-black coloured males. Only at Nkondwe Island do males have light-blue patches. Females are silvery in all the populations known.

Scale-eaters

Perissodus microlepis and *P. eccentricus* both occur in the rocky habitat, but the former is found in all types of habitats and has a lake-wide distribution. There are six different cichlid species in the lake that feed almost exclusively on the scales of other fishes. Scales contain enough nutrition to constitute the major part of a healthy diet (Nshombo *et al.*, 1985).

P. eccentricus received its name because it was found that it has an asymmetric mouth, opening wider on either the lefthand or righthand side. Other investigations have shown that apparently all scale-eaters have an asymmetric mouth (Hori, 1993). Scale-eaters can attack either flank of a victim, but it appears that a scale-eater with a wider mouth opening on the righthand side usually tears scales from the victim's left flank, and vice versa (Hori, 1993). It is clear that the asymmetric mouth is a perfect adaptation for a "hit and run" mode of stealing scales. If an attack is successful the victim is assailed from behind and unaware of what is happening till one or more scales are tugged off its flank. Before it can react the scale-eater has vanished from the scene.

Fish are, however, aware of the possibility of attack, and by the time the majority of the scale-eaters in the area have a left-handed mouth (attacking the right sides of victims), their potential victims will have learned to look out for predators approaching from their right. This means that in such situations scale-eaters which are "right-handed", and approach their victims from the left, have an edge over the majority of their competitors. By the time the population of scale-eaters has slowly shifted to the more favourable right-handedness (because they get more food), the victims are getting accustomed to being attacked more often from the left, so the advantage again shifts the other way. According to Hori (1993) this does indeed take place, and there appears to be,

over time, a continuous switch between left and right in predator and victim.

P. microlepis is the most common scale-eater in the lake and is also recognized immediately as such by other fishes. However, *P. microlepis* attacks its victim from a distance, before it can see the predator coming. *P. microlepis* chooses fishes that are feeding from the algal layer or are displaying frequently; in other words they select victims which are inattentive. The size of the victim usually matches that of the robber. A typical attack starts with a quick dart, sometimes from more than a metre away, followed by a rather violent collision. The impact of the collision usually results in a shower of silvery scales. The victim does not seem to be hurt too much; I have on several occasions seen it just continue browsing from the biocover even when it was displaced by the collision. *P. microlepis* commonly takes advantage of such situations by striking a few times more until the victim eventually flees away.

P. microlepis is a highly specialized cichlid which does not immediately swallow the captured scales, which are neatly stacked before they pass

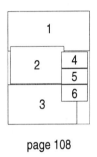

page 108

1. A school of *Asprotilapia leptura* at Samazi, Tanzania.
2. *A. leptura* guarding fry at Isanga, Zambia.
3. *Ophthalmotilapia boops* at Nkondwe Island, Tanzania.
4. A male *O. boops* guardings his territory on top of a large boulder (Nkondwe Island).
O. boops at Kisambala, Tanzania (5) and at Msalaba, Tanzania (6).

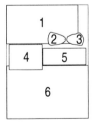

page 109

1. *Perissodus eccentricus* at Ulwile Island, Tanzania.
The asymmetric head of a specimen of *P. eccentricus* seen from the right (2) and from the left side (3).
4. The young of *Perissodus microlepis* seek shelter in the male's mouth (Magara, Burundi).
5. A group of juvenile *P. microlepis* often contains individuals from different pairs.
6. *P. microlepis* guarding its offspring at Kipili, Tanzania.

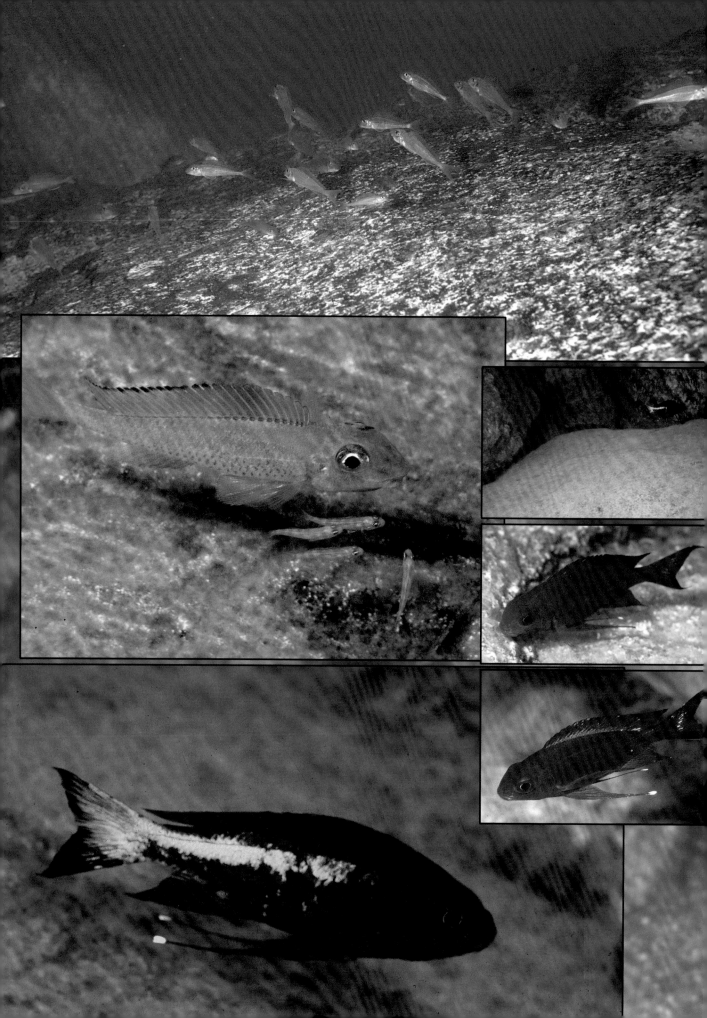

into the digestive tract. After an attack one can see the predator moving its jaws around. Investigation of gut contents revealed perfectly stacked rows of scales in the stomach (Liem & Stewart, 1976). The advantage of stacking is a more efficient filling of the digestive tract. The scales are not macerated by the pharyngeal teeth but slowly digested by the gastric juices. The teeth on the pharyngeal bones are reduced and present only on the edge. The peculiar pickaxe-like teeth in the jaws are bent backwards when the mouth is closed but point forwards when the mouth is extended at maximum aperture (which is further than in most other cichlids). Scales are quickly digested and therefore scale-eaters have a short intestine measuring less than half the fish's total length. The stomach has the shape of an elongated bag and accounts for a third of the tract.

The maximum total length of *P. microlepis* is about 12 cm and that of *P. eccentricus* about 16 cm.

The breeding behaviour of scale-eaters is very interesting as it involves a pair bond even though they are mouthbrooders in which the female broods the eggs. The eggs are fertilized by the male while they are still lying on the substrate. On several occasions it was found that up to 30 minutes elapsed before the female took the eggs into her mouth (Yanagisawa & Nshombo, 1983). Spawns usually number more than 200 tiny eggs. The eggs hatch after two days and after a week the young fry are released by the female. The male continues the task of defending the territory and takes no part in raising the fry until they have reached a length of about 6 mm, which happens on the tenth or eleventh day post-spawning. The male now collects the fry when they are threatened. The chosen territory is usually above a large rock and breeding pairs guarding their offspring are quite common. The fry of *P. microlepis* and *P. eccentricus* eat plankton, as do most young cichlids, but "scale up" their diet when they have attained a length of about 4-5 cm.

Yanagisawa (1985) performed some interesting experiments with brooding *P. microlepis*. In one experiment he removed one of the parents during fry-guarding and found that two things could happen. The most frequent reaction was intensified circling around the batch of fry by the remaining parent. More fry fell prey to predators than when both parents were present. The female turned out to be the best parent. The second possible reaction was the collection of fry by the re-

maining parent, who would then make short scouting trips. When the desperate parent, with its mouth full of fry, encountered another pair of *P. microlepis* with young that were bigger than its own, it would quickly release its own fry among the larger ones. The intruder was soon chased away, but the introduced fry were defended by the foster-parents. The single parent would return to the nest to pick up the next batch of fry and introduce them to another breeding pair.

The breeding behaviour of *P. microlepis* is just a step ahead of the true substrate brooder. The main duty of the male of a substrate brooder is patrolling the territory, while the female takes care of the fry. Later on, when the fry start leaving shelter to feed, the male actively helps in defending them. A lack of shelter (= lack of substrate-brooding sites) may have resulted in the development of mouthbrooding in *Perissodus* and in other species as well. The high number of eggs, their small size, and the early release of fry all suggest that the brooding technique employed by these cichlids is very close to that of substrate brooders. The next step in the evolution of the mouthbrooding process may be biparental mouthbrooding, as observed in *Limnochromis auritus* (see page 258). At a later stage the eggs may increase in size and reduce in numbers, and be incubated for a longer period, as in goby cichlids (see page 27).

P. microlepis occurs in almost every habitat and is not known to have developed geographical variants. *P. eccentricus* was described from specimens caught in a gill-net set at a depth of 100 metres in Chituta Bay, Zambia, in the southern part of the lake, but has also been seen around Kipili in the shallow rocky habitat. It resembles *P. microlepis* in coloration and behaviour but has a deeper body.

The deep rocky habitat

The deep rocky habitat is characterized by sediment-covered rocks and by small patches of mud and sand between the rocks. The subdued light at the depth of this habitat, which is here defined as rocky substrate deeper than 20 metres, prevents most (not all) green algae from growing on the rocks; the majority of the algae covering the rocks are brown to reddish-brown. Non-cichlids, such as catfishes of the genus *Synodontis*, are commonly seen during the day at depths beyond 40 metres. They are common in the shallow rocky habitat as well, but they are nocturnal fishes, leaving their shelter only in the dark.

Neolamprologus buescheri

N. buescheri, named after its discoverer, Heinz Büscher, is a very popular cichlid among aquarists. It was only relatively recently that the first individuals were found at Cape Kachese, the type locality. *N. buescheri* normally occurs at depths greater than about 20 metres, although some specimens may be found in water as shallow as 10 metres (e.g. at Isanga, Zambia). Büscher (1992b) found that the main part of its diet consists of insect larvae (chironomids), but that other invertebrates are taken from the substrate as well.

N. buescheri is a cave brooder and also a cave dweller, rarely venturing more than 30 cm from its cave. Frequently several individuals can be seen together; it seems that sometimes these are pairs and at other times that relatively large juveniles are still "hanging around" the parental cave. In the aquarium *N. buescheri* breeds rather frequently, almost every two weeks, but the batches are very small, sometimes no more than 5 eggs. When the female gets older she produces larger clutches, but the largest spawn I have recorded was 42 eggs. In the wild a male may have several breeding females (usually two) in his territory, which is not very large so the females can be found within a metre of each other. In the aquarium the eggs are attached to the vertical face of a rock, and, probably because there are so few of them, the female almost never fans the eggs. The nest is defended by both male and female, but, as usual, the female stays much closer to the eggs than the male.

As mentioned earlier, the type locality of *N. buescheri* is Cape Kachese in Zambia. Other populations have subsequently been discovered and the overall range of *N. buescheri* currently encompasses the southwestern coast of the Congo, south of Tembwe II (Büscher, 1992b), the entire Zambian coastline of the lake, and into Tanzanian waters where the most northerly population was found at Samazi.

There is some geographical variation in the various populations. The northernmost populations in the Congo have a yellow-brown ground colour and lack the typical dark submarginal band in the dorsal fin (Büscher, 1992b). At Kamakonde the variant is characterized by light-coloured spots in the tail and in the soft-rayed parts of the dorsal and anal fins. At Moliro *N. buescheri* exhibits a

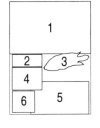

page 112

1. The colourful rocks at Mkinga, Tanzania.
2. A preserved male *Ctenochromis benthicola*.
3. An orange-blotch (OB) female *C. benthicola*. Photo courtesy of Mark Smith.
4. Laif DeMason holding a dried head of *Lates angustifrons*.
5. *L. angustifrons* are common in the deeper rocky habitat (Bulu Point, Tanzania).
6. The eel *Aethiomastacembelus ellipsifer* peeking out of a hole.

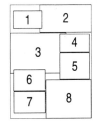

page 113

Neolamprologus buescheri at various localities:
1. Kamakonde, Congo (Aquarium photo).
2. Cape Kachese, Zambia; the type locality.
3. Isanga, Zambia.
4. Chituta, Zambia.
5. Cape Chaitika, Zambia.
6. Samazi, Tanzania.
7. Moliro, Congo.
8. Kantalamba, Tanzania.

pattern of broad vertical bars superimposed on the two horizontal lines. At Cape Kachese the horizontal lines are the most prominent features, together with the blue colour in the fins. At Cape Chaitika the body colour is rather dark and the bars and lines are not prominently visible. Along the western part of Hore Bay ("Gombe"; this is not a real locality name) N. buescheri shows a pattern similar to those found at Moliro and Samazi: the vertical bars are the most prominent. At Chituta N. buescheri exhibits a network of vertical and horizontal elements, while those at Isanga have, in addition to this pattern, many light-coloured spots in the dorsal fin. The latter variant is the most striking when seen in its natural habitat. On the northern side of the Kalambo River the N. buescheri variant resembles that of Cape Chaitika (see photos page 113).

Neolamprologus ventralis and N. bifasciatus

N. ventralis and N. bifasciatus probably have a lake-wide distribution, because both species have been sighted at various places around the lake. N. bifasciatus was first discovered in Zambian waters (Dieckhoff, pers. comm.) and known as "Lamprologus Zambia". Büscher (1993) described this species using specimens collected near Kiku in the Congo. N. bifasciatus has also been recorded from Kigoma, Tanzania (Konings & Dieckhoff, 1992), and along the southwestern coast of the Congo. N. ventralis has not yet been seen in Zambian waters but is known from Kigoma and from Magara in Burundi (Dieckhoff, pers. comm.). Büscher (1995) collected the holotype near Tembwe (locally known as Tembwe Deux) on the Congolese shore.

Both species are rarely found in water shallower than 30 metres and most often they are found at the bottom edge of the rocky habitat where it meets the sandy/muddy substrate. They are most often seen solitary. Aquarium observations suggest that spawning takes place in caves; the spawns are rather small, usually numbering less than 20 eggs.

Geographical variation is not evident when we compare the few specimens observed at the various places around the lake. N. bifasciatus and N. ventralis may both be old species that lack the competitive aggression experienced by other rock-dwellers trying to secure territories in shallower parts of the rocky habitat.

Neolamprologus prochilus

N. prochilus, like the previous two species, lives in rocky habitats deeper than 30 metres. It was described from the population near Mpulungu in Zambia, but Brichard (1978) reports this species from the northwestern coast as well. Thus N. prochilus may, like N. bifasciatus and N. ventralis, have a lake-wide distribution. I have seen it only once in its natural habitat, at Mbete Island near Mpulungu. Although it in some ways resembles Altolamprologus compressiceps, it is much larger than the average-sized compressiceps and lives a much more secluded life, remaining close to the substrate inside large caves. N. prochilus is easily identified by its huge mouth, the jaw bones of which protrude noticeably from the head. The mouth is very narrow and can be extended far forward from the rest of the head, facilitating the capture of prey. The diet consists of invertebrates which are located in the dark recesses of the deep rocky habitat. The sensory system in N. prochilus is very much enhanced and large pores are visible on the head. These, of course, are needed for the detection of prey in the dark. In contrast to A. compressiceps and A. calvus, the scales of N. prochilus are not thick and hard and N. prochilus must thus be more vulnerable to bruises and scratches from collisions with rocks.

Brichard (1989) reports that N. prochilus is a cave brooder that forms families. He found pairs defending fry that had a size of 2 to 3 cm. This differs from the behaviour of compressiceps and calvus in that the males of these two species leave most of the guarding duties to the female, and fry of 2 to 3 cm are very rarely seen together with the female. Brichard further reports that spawns are rarely larger than 50 eggs and a maximum of 20 large fry have been found guarded by a pair.

Neolamprologus variostigma

N. variostigma is a recently described taxon (Büscher, 1995) which was collected at a depth of 45 metres on the Congolese shore near Tembwe (Tembwe Deux). It appears to be very rare as it was seen only twice during many surveys undertaken by Büscher. The species is characterized by having a very short upper lateral line consisting of no more than 4 to 5 pored scales which are separated by scales without pores. Another peculiarity of N. variostigma is the very short pectoral fins (Büscher, 1995). This species was not investigated by Stiassny (1997) so its placement in *Neolamprologus* may be inaccurate; it may, after appropriate research, need to be transferred to *Lepidiolamprologus* or to a genus of its own.

Elongate predators

The genus *Lepidiolamprologus* is represented by several species in the deep rocky habitat but a single species, *L. profundicola*, is the most prevalent. It is a large pursuit predator — maximum total length 31 cm — which hunts small fishes by chasing them. It has a lake-wide distribution but geographical variation is unknown. Breeding pairs or guarding females are most often seen in the shallow rocky and intermediate habitats. The eggs are attached to the vertical face of a rock and guarded by both male and female. Although most predators are kept at bay the pair overlook the tiny egg-robbing *Telmatochromis* species, which are frequently seen feeding on the eggs while the parents hover over the nest. The male probably leaves the female — he may have a harem, not uncommon among lamprologines — before the fry swim free. A clutch may number more than a thousand eggs but many of them are eaten before they hatch.

Near Kipili (Dieckhoff, pers. comm.) and around Fulwe Rocks, near Wampembe on the Tanzanian shore of the lake, a species has been discovered that resembles both *L. profundicola* and *L. elongatus*. Interestingly this undescribed species, *L.* sp. "profundicola tanzania", is found sympatrically with *L. profundicola* and with *L. elongatus*. The latter two species differ in the size of the eye — that of *L. profundicola* is, relatively, much smaller than that of *L. elongatus*. The relative eye size of "Profundicola Tanzania" is intermediate between that of the other two although it tends to be closer to that of *L. elongatus*. The undescribed species differs from the other two in coloration by virtue of the yellow anal fin of subadult specimens and the vertical barring in breeding individuals. The estimated total length of *L.* sp. "profundicola tanzania" is about 18 cm.

A few fry-guarding females were seen at the base of the rocks at a depth of about 22 metres. *L. elongatus* almost always has its nest in the pure rocky habitat, but *L. profundicola* and the undescribed species seem to have their nests in intermediate types of habitat.

Cyphotilapia frontosa and its mimic

Cyphotilapia frontosa is a common species in the deeper rocky habitat and lives in areas deeper than 15 metres. The average depth of occurrence depends on the population and habitat: at Milima Island, one of the Kavala group, large numbers of *C. frontosa* are found at depths ranging between 5 and 40 metres; at Kapampa the shallowest places where one can encounter *C. frontosa* are at depths between 25 and 30 metres. This majestic species is found in every habitat with rocks at a depth of 20 metres and deeper. *C. frontosa* lives in groups, and it is not uncommon to find a large male, who is in charge of the group, several females, sometimes brooding, and one or two smaller males. Juveniles can be found at somewhat shallower levels, but adults, which can attain more than 35 cm in total length, are observed mainly below 20 metres.

C. frontosa is quite lethargic and never seems to be in a hurry unless it is chased by divers, but this energy-saving strategy does not exempt it from

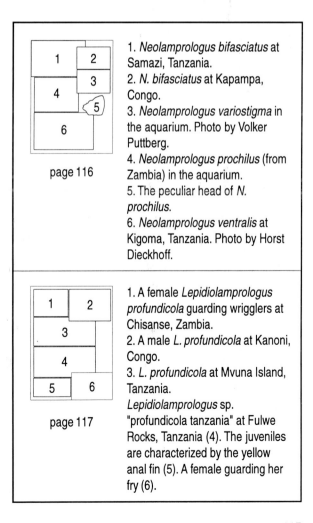

page 116

1. *Neolamprologus bifasciatus* at Samazi, Tanzania.
2. *N. bifasciatus* at Kapampa, Congo.
3. *Neolamprologus variostigma* in the aquarium. Photo by Volker Puttberg.
4. *Neolamprologus prochilus* (from Zambia) in the aquarium.
5. The peculiar head of *N. prochilus*.
6. *Neolamprologus ventralis* at Kigoma, Tanzania. Photo by Horst Dieckhoff.

page 117

1. A female *Lepidiolamprologus profundicola* guarding wrigglers at Chisanse, Zambia.
2. A male *L. profundicola* at Kanoni, Congo.
3. *L. profundicola* at Mvuna Island, Tanzania. *Lepidiolamprologus* sp. "profundicola tanzania" at Fulwe Rocks, Tanzania (4). The juveniles are characterized by the yellow anal fin (5). A female guarding her fry (6).

the need for its daily dinner. Young *C. frontosa* feed on soft-bodied crustaceans but (sub)adults are piscivores. The teeth on the pharyngeal bone are sharp and slender in shape, in strong contrast to the molariform dentition found in macro-invertebrate feeders. Stomach contents of wild specimens have revealed mostly remains of fish (Poll, 1956). Observations in captivity add credibility to the idea that *C. frontosa* is a piscivore: it can — and does — devour tankmates that are almost half its own size.

In nature the main source of food for *C. frontosa* can usually be found swimming just above the predator: cichlids of the genus *Cyprichromis*. Immense schools, numbering more than tens of thousands of *Cyprichromis*, provide a nice "restaurant" for many *C. frontosa*. Of course, the frontosa needs to use subterfuge to catch its dinner, because it is not going to expend energy in a pursuit chase in which it will definitely lose out to the agile *Cyprichromis*. However, at dusk most cichlids become inactive, and the *Cyprichromis*, which spend the day in the water column, spend the night resting on the substrate. During the twilight period, while other cichlids are becoming lethargic, or, in the morning, are "coming awake", *C. frontosa* is up and about and getting ready for a meal. The almost unconscious *Cyprichromis* are shovelled up with hardly any resistance and many end their days stuffing the stomach of this sneaky predator. During the day *C. frontosa* is found lurking beneath *Cyprichromis* schools and may feed on "geriatric" male *Cyprichromis* that have lost their territory to stronger, younger, ones.

The nuchal hump is most prominent in males but females are not completely devoid of such a gibbosity. The hump is formed by the dorsal muscle that tends to extend forward and may play a role in mate recognition. *C. frontosa* breeds in a peculiar manner not known from any other cichlid. Before spawning takes place the blue colour of a male intensifies; in particular on the snout. Males do not dig spawning sites or defend territories (too much effort!). When a male and female get ready to spawn they do not separate from the group but look for a suitable site which is then weakly defended by the male.

The male initiates spawning by discharging his seminal fluid in the centre of the nest, even before the female has laid a single egg. He does not quiver or undulate his body, as is seen in most other mouthbrooding species, but slowly moves over the site, with fins folded, showing the female where to go. After the male has staged this demonstration run the female moves in a similar fashion over the site. However, she does not swim in circles like other cichlids, but moves backwards (!) to pick up the egg(s) just laid. Having done so she does not make way for the male but continues to deposit eggs. If undisturbed the female can "swing" five or six times in a row, having deposited perhaps up to twenty eggs, before the male enters the nest again. In all probability the male's milt has enough "power" to fertilize the eggs even though it is discharged several minutes before it actually contacts them. The female has never been seen to nuzzle the male's anal fin, a common practice in other mouthbrooders. After the female has left the site the male again leads her back, and the whole process is repeated until all the eggs have been laid. The male seems never to chase the female. The eggs are brooded for about five weeks before the fry are released.

C. frontosa has a lake-wide distribution and occurs in several distinct races. Initially only the populations from Burundi were exploited for the aquarium hobby. Then the Kigoma race was introduced. The latter variant has one more vertical band on the body than the Burundi race and has thus become known as the "Seven-Band Frontosa", but in view of the other variants subsequently discovered it is better to refer to it as *C. frontosa* (Kigoma) or "Kigoma Frontosa". It is characterized mainly by the dark blue colour on the cheeks and the variable yellow coloration in the dorsal fin. The third variant exported from the lake comes from Zambian waters and is characterized by a brighter blue coloration (in comparison with the northern races) and an interorbital band. In the late eighties a very dense population of *C. frontosa* was discovered near the Kavala archipelago in the Congo. This variant resembles the Burundi race but occurs in rather shallow water (up to 5 metres).

In 1990 a new variant of the frontosa was discovered in the Congo, where it is distributed along the rocky shores between M'toto and Kapampa. This population is characterized by the intense blue colour on all fins and the mother of pearl on the hump and upper part of the body in large specimens. Juveniles also show the attractive blue coloration seen in adults. It occurs, however, at deep to very deep levels. The shallowest point at which I have seen this blue frontosa was

25 metres. Most specimens are collected at a depth of between 40 and 60 metres and require three days to decompress.

The latest geographical variant to be exported hails from Cape Mpimbwe, Tanzania, and is characterized by large pearl-coloured scales on the humps of large males.

An interesting phenomenon is the occurrence of a scale-eating cichlid, *Plecodus straeleni*, which has an identical coloration to that of *C. frontosa*. It has been reported (Brichard, 1978, 1989) that *P. straeleni* uses its coloration as camouflage to enable it to mix with groups of *C. frontosa* and stealthily scrape some scales from the flanks of the unsuspecting prey. I have never observed *P. straeleni* attacking *C. frontosa* although I have seen them among their schools. However, I have observed *P. straeleni* attacking other cichlids, such as *Cyathopharynx furcifer*, or even the much smaller *Neolamprologus brichardi*.

It has also been said that it may mimic not only *C. frontosa* but also *Neolamprologus sexfasciatus* and *N. tretocephalus*. There are several reasons why this seems unlikely. First of all *P. straeleni* resembles *C. frontosa* to such an extent that it is easily mistaken for its model, even at close quarters. It goes to great lengths to copy *C. frontosa*, so much so that the population at Kapampa has a much bluer coloration (like *C. frontosa*) than that at Rutunga, Burundi (see photos page 121). Secondly, there are no *N. sexfasciatus* or *N. tretocephalus* in Burundi waters, which does not, of course, preclude the possibility that *P. straeleni*'s colour pattern might have developed in mimicry of those species, but would still render inexplicable its extreme resemblance to *C. frontosa* at locations where these two species are found. Thirdly, we could argue that other scale-eaters, especially the abundantly present *Perissodus microlepis*, do not mimic their prey. If we combine this with the observation that *P. straeleni* attacks other cichlids than frontosa then it seems unlikely that *P. straeleni* needs its coloration in order to be able to exist predominantly on scales of *C. frontosa*.

There is a far more plausible explanation. *C. frontosa* roams about in the rocky biotope but has never been seen hunting prey. Other species inhabiting the same biotope are probably aware of its peaceful manners and do not expect to be attacked, their complacency allowing *P. straeleni*, "the wolf in sheep's clothing", its opportunity. Feigning to be a good-natured *C. frontosa*, the scale-eater can easily close in on its prey, and before the victim can recognize the wolf for what it is, it has lost some of its scales.

The maximum total length of *P. straeleni* is about 14 cm. It usually occurs in somewhat shallower water than *C. frontosa*, and, to confuse the issue, it can have two completely different colour patterns: the barred pattern that mimics the frontosa, or a light-brown to black coloration. The light-coloured individuals are females and the dark ones are males (Yanagisawa *et al.*, 1990). Nshombo (1991) found that females often steal newly-laid eggs of mouthbrooding cichlids instead of scales. Maybe this stimulates the production of their own eggs.

Breeding *P. straeleni* are very dark brown but have rows of dark blue scales. Like all other Tanganyikan scale-eaters they are biparental mouthbrooders. The female broods the eggs and larvae, but after the fry have first been released they are cared for by both parents for a long period and may shelter inside their parents' mouths for more than a month after release.

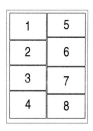

Cyphotilapia frontosa at various localities around the lake:
1. Magara, Burundi.
2. Milima Island, Congo.
3. Kapampa, Congo.
4. Cape Chaitika, Zambia.
5. Kigoma, Tanzania (aquarium).
6. Bulu Point, Tanzania.
7. Msalaba, Tanzania.
8. Kantalamba, Tanzania.

page 120

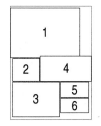

1. *Cyphotilapia frontosa* at Fulwe Rocks, Tanzania.
2. The peculiar teeth of the scale eater *Plecodus straeleni*.
3. *P. straeleni* guarding fry at Msalaba, Tanzania.
4. A blue-coloured *P. straeleni* at Kapampa, Congo.
5. A light-coloured *P. straeleni* at Magara, Burundi.
6. A female *P. straeleni* at Kantalamba, Tanzania.

page 121

Tanganyika butterflies

Most members of the genus *Xenotilapia* are sand-dwelling species which filter the sandy substrate through their gills in search of something edible. There are, however, a few *Xenotilapia* species found in the rocky habitat that feed from the layer of sediment covering the rocks. They are never found in large groups or foraging schools like their sand-dwelling cousins, but occur in pairs or small groups of juveniles. The species living in the rocky habitat are *X. papilio*, *X.* sp. "papilio katete", and *X.* sp. "papilio sunflower".

X. papilio was described by Büscher (1990) using a specimen collected near Tembwe (Tembwe Deux) in the Congo. He found this species at a depth of only 3 metres, but when other populations were found it became apparent that *X. papilio* is found mostly in the deeper rocky habitat, in particular where the rocks are covered with sediment. Their food consists of invertebrates which are extracted by filtering the sediment through the gills.

X. papilio and the other two, undescribed, species belong to the biparental mouthbrooders of the lake. The pair defend a territory on the upper face of a large rock and seem to stay together in this area even when not breeding. In the aquarium spawning takes place at a specially selected but not prepared site. Clutch sizes are rather small, which is to be expected of such a small fish (maximum total length about 10 cm), and hardly ever number more than 15 eggs. The female initially broods the eggs and larvae and after about 9 to 13 days the latter are transferred to the male's mouth. Three weeks after spawning the fry are released but still guarded by both parents. When threatened the fry find shelter in both parents' mouths for at least another two weeks, but not all the fry will fit into the two mouths during the last week of parental care.

X. sp. "papilio katete" resembles *X. papilio* in many ways and is distinguished only by a different morphology: it has a deeper body than *X. papilio*, and light-blue to clear pelvic fins whereas the latter are yellow in *X. papilio* and *X.* sp. "papilio sunflower". The "Sunflower" has a shallower body than *X. papilio* but the difference is very small. The most noticeable difference is the black spots in the yellow pelvic fins of *X. papilio*; these markings are absent in the Sunflower.

Even though *X. papilio* has the smallest range of these three species, two distinct geographical variants are known. *X. papilio* occurs along the coast between Kanoni and Tembwe. At Kanoni the adults have yellow dorsal fins without black markings; at Tembwe, however, the dorsal is vividly marked with black and white spots. *X.* sp. "papilio katete" is found between Kapampa in the Congo and Katete in Zambia. There is no obvious geographical variation when specimens from three different localities are compared (see photos page 124).

X. sp. "papilio sunflower" has the widest range and occurs between Chituta in Zambia and Cape Mpimbwe in Tanzania. The form found at Chituta lacks the large black spot(s) in the dorsal, but has many tiny spots. There is very little variation throughout its entire range along the east coast, and although the spot in the dorsal tends to be round, instead of elongate, in the northern part of the range, there are also specimens in the north that have the elongated blotch(es) commonly seen in the southern populations.

Greenwoodochromis

Greenwoodochromis christyi belongs to a group of Tanganyika cichlids, members of a number of different genera, which have a colour pattern consisting of a few horizontal rows of iridescent scales and a small number of broad vertical bars. Probably the best-known member of this group is *Limnochromis auritus* (see page 258), while the other species involved are *L. abeelei*, *L. staneri*, *Greenwoodochromis christyi*, *G. bellcrossi*, and *Gnathochromis permaxillaris*. The main difference between the species transferred to *Greenwoodochromis* and those remaining in *Limnochromis* is that the former have more than 48 scales in an horizontal row between the gill-cover and the tail, whereas the maximum in *Limnochromis* is 40. Given the tendency of some modern revisions of cichlids (e.g. Malawi cichlids: Eccles & Trewavas, 1989) to distinguish genera mainly by their basic colour pattern, it would not surprise me if all these species were returned to their original generic assignments in the future.

The two species currently in *Greenwoodochromis* are distinguished from each other by the size of the eye and the shape of the mouth. *G. christyi* has a relatively small eye compared to that of *G. bellcrossi*. The latter also has a steeply inclined mouth, while that of *G. christyi* is only moderately inclined. *G.bellcrossi* has been confused with female *Hemibates stenosoma* (the photo on page 241 of Konings, 1988, and on page 215 of Herrmann,1987 & 1990, depicts a female *H. stenosoma* rather than *G. bellcrossi*) but can be distinguished from such females by its moderately truncate tail — that of *H. stenosoma* is forked. It also has fewer soft rays in the dorsal: 9-11 versus 13-15 in *H. stenosoma*. Male *H. stenosoma* (see page 133) are readily identified by the black markings on the body.

G. christyi is seen in the deeper rocky habitat and has been found mainly in the southeastern corner of the lake; it has, however, also been caught near Uvira in the Congo (De Vos *et al.*, 1996). I have seen this species at Chituta, Zambia, and near Samazi, Tanzania. *G. christyi* is a biparental mouthbrooder, with both female and male taking care of the offspring. Both male and female brooding the larvae, and leading the fry, once released, through the habitat, seems to be characteristic of Limnochromis-like species, as this behaviour is also found in *L. auritus* and *Gnathochromis permaxillaris*. Broods number between 50 and 250 fry.

The vacuum cleaner

One of the most remarkable cichlids of Lake Tanganyika is *Gnathochromis permaxillaris*. Its most interesting features are the very large mouth, which opens in a quite unique way (see photos page 128), and the shape of the upper lip. It is a fairly large cichlid which is known to attain a total length of more than 18 cm.

G. permaxillaris lives on the muddy patches between the rocks at depths of more than 30 metres and also on the muddy bottom at the edge of rocky habitats. Although the enormous mouth may suggest otherwise, it feeds on the very small organisms found over muddy substrates: tiny invertebrates that live close to the bottom, hovering a few millimetres above the substrate. The wide gape of permaxillaris spreads the flow of water into its mouth, when the gill covers are opened, over a relatively wide area. This slows the flow so that only small light particles are carried into the mouth. This cichlid thus effectively "vacuums" the muddy bottom, and has to do so continuously in order to obtain sufficient food.

Its foraging grounds may extend into the anoxic layers below 200 metres, as some specimens have

been collected at such depths (Poll, 1956). Poll also found that most specimens collected fell into three categories, indicating that it takes about three years for *G. permaxillaris* to mature, and that breeding takes place during a particular season. The first year the juveniles grow to a length of about 7 cm, the second year up to 11 cm, and during the third year they mature and breed.

Breeding takes place in holes dug in the mud or sand under rocks. Both male and female brood the eggs, although for the first day the female broods the eggs on her own. After that the eggs and larvae are exchanged between the partners several times a day. The eggs hatch after five days and the fry are released about 12 days after spawning (Eysel, 1992). Broods number between 30 and 100 fry.

G. permaxillaris is found throughout the lake but never in large numbers. Males and females appear to have a similar coloration, but the female seems to be slightly smaller than the male. *G. permaxillaris* have longitudinal rows of yellow-coloured scales on the flank, and these are more prominent in the Burundi population, but in general individuals of all known populations have a very similar colour pattern.

Ctenochromis benthicola

A few specimens of *Ctenochromis benthicola* have been caught at the northern end of the lake, but only very rarely has this species been observed in its natural environment (Brichard, 1979). All specimens were caught or seen at a depth of 30 metres or deeper. The males caught up till now have measured between 10 and 15 cm, whereas the maximum recorded total length for females is

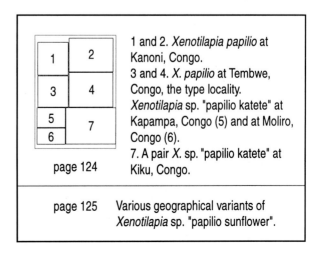

1 and 2. *Xenotilapia papilio* at Kanoni, Congo.
3 and 4. *X. papilio* at Tembwe, Congo, the type locality.
Xenotilapia sp. "papilio katete" at Kapampa, Congo (5) and at Moliro, Congo (6).
7. A pair *X.* sp. "papilio katete" at Kiku, Congo.

page 125   Various geographical variants of *Xenotilapia* sp. "papilio sunflower".

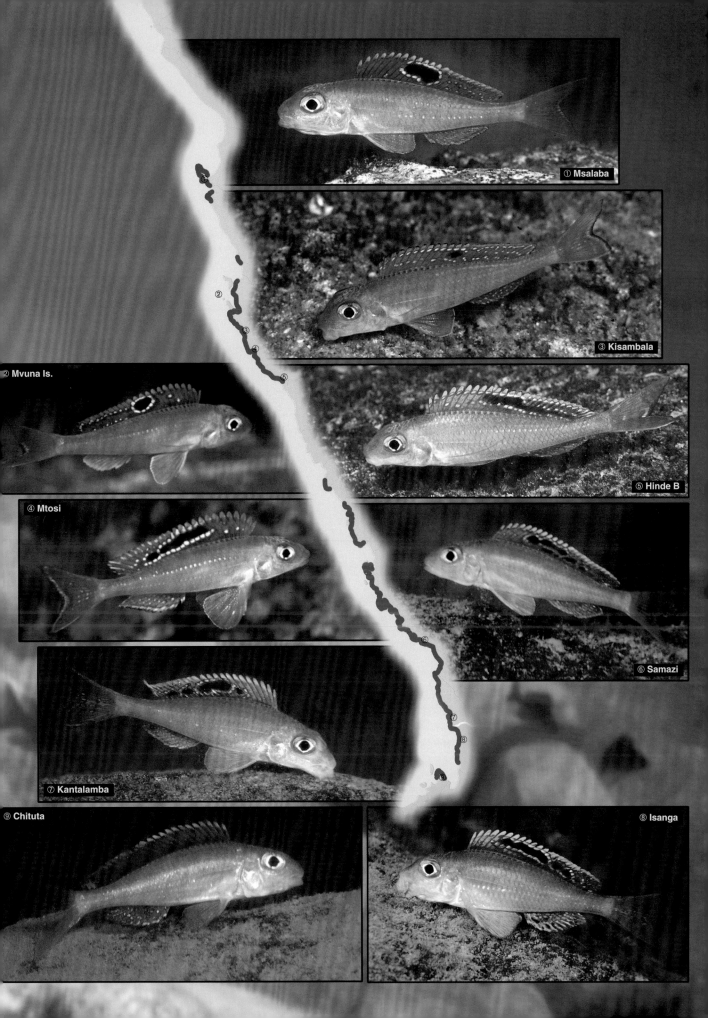

① Msalaba

③ Kisambala

② Mvuna Is.

⑤ Hinde B

④ Mtosi

⑥ Samazi

⑦ Kantalamba

⑨ Chituta

⑧ Isanga

a substantial 22 cm! The male is a dark, blue-brown, colour, in marked contrast to some females which are bright orange (see photo page 112). Brown females have also been caught (Mireille Schreyen, pers. comm.). Only a few females have been caught so far; males are seen more often. *C. benthicola* has enlarged sensory pores on the head, a feature sometimes found in other species living in dark and/or deep regions of the lake.

The depths

A description of this habitat is difficult to give as nobody has yet seen it. The species discussed here have rarely been collected in water shallower than 50 metres, and we know about them only from fishermen who have caught them either by hook and line or with deep-set gill nets. Most of these species probably live in close contact with the bottom substrate, which consists mainly of a muddy substance (Poll, 1956). Because hardly any light penetrates to these depths, epilithic algae-eaters (those feeding from growths on rocks) are not found here. The rocks are probably largely covered with a thick layer of muddy sediment, and hence caves are rare. The lack of light has probably stimulated the development of non-visual sensory systems for detecting food.

Sonar-feeding cichlids

The members of the genera *Trematocara* and *Telotrematocara* are specialized in feeding in the dark, or at least under minimal light conditions. Poll (1956) reports that most species are abundant and gregarious, and have a lake-wide range, but the recently described *T. zebra* is known only from Luhanga and Pemba in the northwestern part of the lake (De Vos *et al.*, 1996). When the sun sets *Trematocara* species migrate to shallower water, and can sometimes be found in water as shallow as a few metres. The same species, and probably the very same individuals, can be found at a depth of 200 metres during the day.

This remarkable vertical migration is striking by virtue of the fact that the fish endures an enormous variation in pressure, and the more so when we consider that the swim bladder is closed. Under normal circumstances a cichlid living in the upper 30 metres of water can move around freely in a layer five metres thick without impairing its buoyancy. Larger vertical movements are, however, another matter. Cichlids have a closed swim bladder and thus cannot adjust their buoyancy instantaneously by releasing some of the gases in the bladder. In most species the process takes time — on average a cichlid can adapt to a difference of five metres of water pressure per 24 hours — and if one is caught at a depth of, for example, 100 metres and immediately hauled to the surface, it does not have enough time to adjust the gas content of its swim bladder to the much lower pressure of the shallow water. As a result the bladder expands and squashes the other internal organs, and often parts of the gut are forced out of the body cavity, via either the anal opening or the mouth. This process is irreversible, and the fish dies within a few minutes if not already dead before it is hauled into the boat. And by the same token, vertical migrations in excess of the tolerated range will produce a bulging or a compressed fish unable to swim properly.

This is in strong contrast to the capabilities of the species of the genus *Trematocara* and some other genera to be discussed later. Their specialized swim bladder allows these fishes to adjust rapidly to changing water pressures during their daily migrations. In this way they can feed in the food-rich upper regions of the lake without any fear of being eaten by diurnal piscivores, and they have no need to compete for breeding space with more aggressive, or simply larger, species.

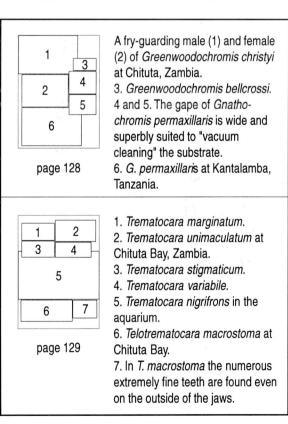

page 128

A fry-guarding male (1) and female (2) of *Greenwoodochromis christyi* at Chituta, Zambia.
3. *Greenwoodochromis bellcrossi*.
4 and 5. The gape of *Gnathochromis permaxillaris* is wide and superbly suited to "vacuum cleaning" the substrate.
6. *G. permaxillaris* at Kantalamba, Tanzania.

page 129

1. *Trematocara marginatum*.
2. *Trematocara unimaculatum* at Chituta Bay, Zambia.
3. *Trematocara stigmaticum*.
4. *Trematocara variabile*.
5. *Trematocara nigrifrons* in the aquarium.
6. *Telotrematocara macrostoma* at Chituta Bay.
7. In *T. macrostoma* the numerous extremely fine teeth are found even on the outside of the jaws.

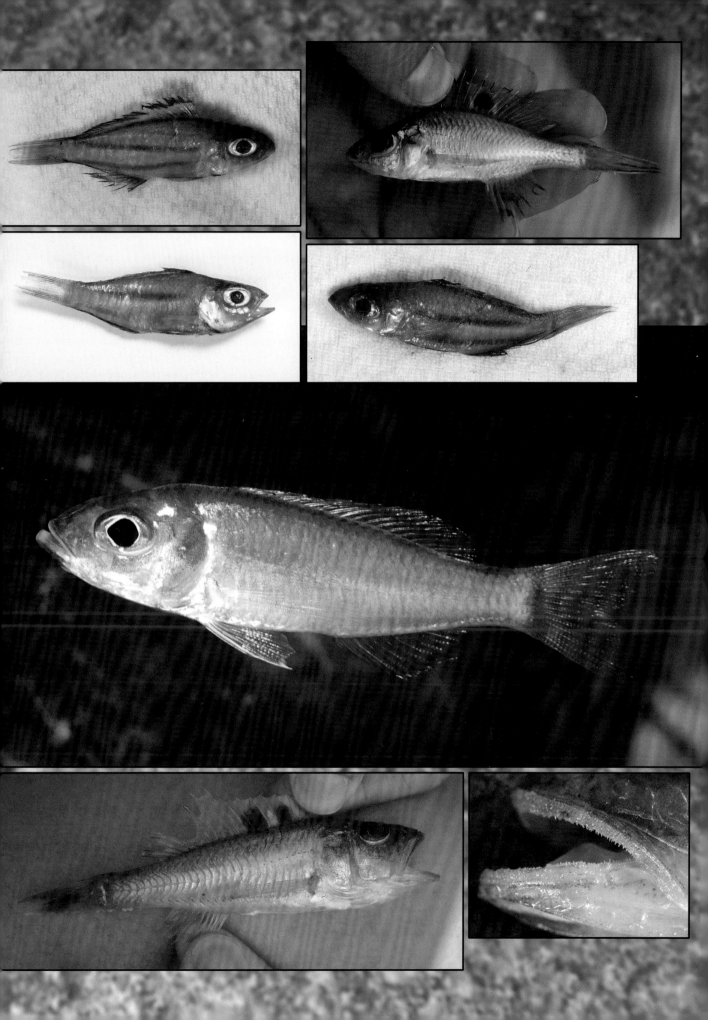

In order to locate food in the dim light, *Trematocara* species have developed an enhanced acoustic system (the sensory pores of the head and lateral line) around the mouth. This sensory system has a similar function to that of our ears, and registers pressure waves in the water originating from moving objects. In many species the acoustic system serves to warn of approaching predators, while in others it plays a role in their schooling behaviour. In *Trematocara* the enhanced sensory system on the head functions as a prey detection system, enabling the fish to detect prey (e.g. swimming or breathing invertebrates) that cannot be seen because of the poor light conditions prevailing at feeding time — during the night many insect larvae and crustaceans leave their shelter in order to feed on the plankton, and these form an easy prey for *Trematocara*. And by virtue of this sensory system and the specialised swim bladder, *Trematocara* species can compete effectively with other species for the same food without becoming embroiled in territorial battles or even being seen by these other species.

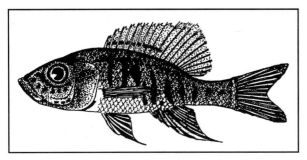

Trematocara zebra (from De Vos, Nshombo, Thys van den Audenaerde, 1996)

The extensive sensory system on the head of *Trematocara* may also have allowed it to reduce the acoustic system (lateral line system) on the rest of its body: in all species the lateral line is considerably reduced or even absent.

There is some food differentiation among the five species that feed on the above-mentioned invertebrates. *T. unimaculatum* feeds on small snails as well as on insect larvae and crustaceans (Poll, 1956). It is possible that such snails (common in muddy substrates) are taken during the day while the fish is in deep water, but *T. unimaculatum* may just sift the substrate for anything edible and retain hard-shelled prey as well. The pharyngeal teeth are strong and enlarged on the centre of the

bones, indicating a hard-shelled food. *T. variabile* is specialized on insect larvae but may feed on crustaceans as well. *T. nigrifrons* and *T. stigmaticum* eat anything that moves and is small enough to be swallowed. It is not uncommon to find an occasional small cichlid in the stomach of *Trematocara* species (Poll, 1956). *T. marginatum* does not seem to be specialized on any particular source of food, and feeds on any small soft prey. The small *T. zebra* seems to favour shrimps (De Vos *et al.*, 1996).

Breeding probably takes place in deep water, as it has never been observed in the shallows and mouthbrooding females have been trapped in gill nets set at great depths (Poll, 1956). It is not known whether these species are lek (arena) breeders in which males defend small territories clustered in large groups, but this would not be uncommon for a gregarious mouthbrooding species. Moreover schooling species often breed and release their fry simultaneously, with the fry forming a new school and probably staying together for their entire life. The only record of breeding *Trematocara* in captivity is that of *T. nigrifrons* by Krüter (1991), who found that the female is noticeably larger than the male and that the male exhibits black and white edges to the fins and a black throat when breeding. Spawning was not observed, but the female brooded for three weeks, producing about 40 fry which had a length of 8 mm when released.

Males are distinguished from females by their specific black markings, and each species has its own specific pattern. Since several species are often found together throughout the lake they need to have different patterns in order to enable the female to select the correct mate.

Males of *T. unimaculatum* have a large black spot on the dorsal, and this marking is much smaller in the female. Interestingly females appear to be larger than males and attain a length of approximately 12 cm — the average length of males is 8.5 cm. The maximum total length (female) is 15 cm, and this species is the largest of its genus. It is possible that this species has different breeding behaviour to the others, this hypothesis being based not only on the larger dimensions of the female but also on the fact that both her ovaries are functional: biparental mouthbrooding is commonly found in mouthbrooders with two active ovaries. A large number of eggs seems to be a prerequisite for a substrate-brooding species devel-

130

oping a mouthbrooding reproduction mode. The intermediate step usually involves biparental care since large numbers of eggs cannot be brooded by the female for a long time. During such development egg sizes increase but batch sizes decrease and are handled by the female alone. The final step is the maternal mouthbrooding mechanism in which small batches of very large fry can withstand the stress of predation after being released.

T. variabile has a maximum length of 8.5 cm. Males are distinguished by their large dorsal fin and two horizontal stripes on the sides, which are absent in females. Males of *T. nigrifrons* have black pelvic fins and gill cover (branchiostegal) membranes, while these are clear in the females — females of this species are also larger than males. Maximum size is 11.5 cm. *T. marginatum* is characterized by a dark marginal band in the unpaired fins in both male and female. Males can be readily identified by the two horizontal stripes on the body which are absent in the female. The maximum total length this cichlid can attain is 10 cm. *T. stigmaticum*, which may live in shallower water than the other species (Poll, 1956), exhibits a black submarginal bar in the dorsal fin of the male, whose pelvic fins become black when he is motivated. *T. zebra* (maximum total length approximately 8 cm) males exhibit a pattern of thin vertical bars and a black stripe in the spiny part of the dorsal fin. In females the vertical bars are reduced to a few spots and are obscured by a wide horizontal band extending from the edge of the gill cover for about the length of the pectoral fin.

Telotrematocara macrostoma was described from two juvenile specimens, and was originally assigned to *Trematocara*. Poll (1986) argued that the teeth in *macrostoma*, which are very small and numerous and also found on the outer surface of the jaws, merit its isolation in a monotypic genus, *Telotrematocara*. However, De Vos *et al.* (1996) found that external teeth also occur in other *Trematocara* species and thus should not be the sole basis for separating *macrostoma* from the other trematocarini.

T. macrostoma is further characterized by two ocellated spots in the dorsal fin and by its large mouth, which is unequalled by that of any other cichlid and may function as a kind of basket for collecting plankton as the fish swims with its mouth open. However, although several adult specimens have been collected none of them had food remains in its stomach. Bailey and Stewart (1977) also discuss this species and mention that males have a much higher dorsal fin than females, but again no mention is made of its possible diet. The sensory system on the head is well-developed. Maximum total length is approximately 10 cm. All specimens were caught in the southern half of the lake together with large quantities of six species of the genus *Trematocara*; *T. macrostoma* is, however, not an abundant species.

Scale-eaters from the depths

Plecodus multidentatus may feed on the scales of *Trematocara* species. This scale-eater attains a maximum length of 12 cm and has a silvery body. The lyre-shaped tail has four vertical bars. The shape of the teeth is like that of a pick-axe, as seen in other scale-eaters. Stomachs have been found to be crammed with neatly-stacked scales. Nothing is known about its breeding behaviour although the largest specimens collected were females (Poll, 1956), one of which had about 44 fully developed eggs, with a length of 2.5 mm, in her ovaries. The type locality of *P. multidentatus* is near Moba, Congo, but specimens have been caught all around the lake.

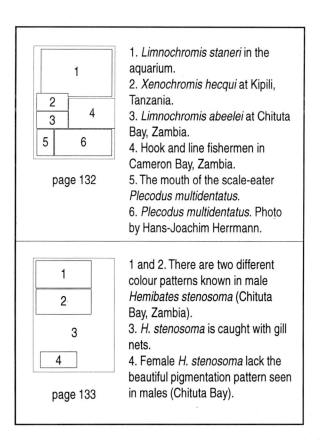

1. *Limnochromis staneri* in the aquarium.
2. *Xenochromis hecqui* at Kipili, Tanzania.
3. *Limnochromis abeelei* at Chituta Bay, Zambia.
4. Hook and line fishermen in Cameron Bay, Zambia.
5. The mouth of the scale-eater *Plecodus multidentatus*.
6. *Plecodus multidentatus*. Photo by Hans-Joachim Herrmann.

page 132

1 and 2. There are two different colour patterns known in male *Hemibates stenosoma* (Chituta Bay, Zambia).
3. *H. stenosoma* is caught with gill nets.
4. Female *H. stenosoma* lack the beautiful pigmentation pattern seen in males (Chituta Bay).

page 133

Plecodus elaviae is a scale-eater which has been caught in gill nets at an average depth of 50 metres (Poll, 1956). This species is known from a few localities in the northern half of the lake. A large group (23 specimens) was caught near Kalemie at a depth of 70 to 100 metres. This species seems to feed exclusively on scales because it is unknown to fishermen who catch fish with hook and line.

P. elaviae is a much larger scale-eater — it attains a maximum total length of 32 cm — and it seems likely, as has been observed in other scale-eaters, that its prey has similar dimensions. *Bathybates* species and some other deep-living species may therefore be its main target.

Xenochromis hecqui — maximum total length about 29 cm — may also have specialized on larger prey such as *Bathybates* species. It is characterized by two dark spots on the dorsal part of the flank above the anal fin, and extending into the dorsal fin. As is the case with most cichlids from the depths nothing is known of its breeding technique. Poll (1956) mentions that the largest specimen collected was a female and that her ovaries contained about 600 eggs with an average length of about 2.5 mm. *X. hecqui* has a lake-wide distribution.

Limnochromis

Limnochromis abeelei and *L. staneri* are rare species which nevertheless may have a lake-wide distribution. The holotypes of both species were collected near the Kavala Islands in the Congo, but other specimens of *L. abeelei* have been collected along the central part of the Congolese shore and near Karema in Tanzania (Poll, 1956); *L. staneri* is also known from other localities in the central part of the lake (Poll, 1956). Both species have also been collected in Zambian waters (Bailey & Stewart, 1977). *L. staneri* resembles *L. abeelei* in many details of its morphology but the main differences are the molariform teeth in the pharyngeal jaws and the deeper body (2.8 - 3 times in standard length versus 3.3 - 3.5 in *L. abeelei*). *L. abeelei* also attains a larger size: 24 cm as opposed to 19 cm in *L. staneri*. This is important, as a difference in pharyngeal tooth shape alone is not a very convincing character for distinguishing two sympatric species. There are several different cichlid species known where the shape of the pharyngeal teeth varies among individuals of the same population (e.g. in *Aulonocara* species from Lake Malawi and in *Herichthys minckleyi* and *H. labridens* from Mexico).

L. abeelei feeds mainly on fish and large crustaceans whereas stomach inventories of *L. staneri* have revealed remains of snails in addition to fish and shrimps (Poll, 1956). The thick pharyngeal teeth of *L. staneri* enable it to crush small snail shells whereas *L. abeelei* is restricted to softer prey.

Little is known about their breeding behaviour except that both are mouthbrooders which may have a similar breeding mechanism to the biparental mouthbrooder *Limnochromis auritus*. Poll (1956) reports that some sexual difference can be found in the length of the filaments of the dorsal, anal, and ventral fins, which are slightly longer in males. He found 200 eggs with a maximum length of 2 mm in one of the ovaries of a female *L. staneri*.

The rarest Tanganyika cichlid

Baileychromis centropomoides, whose shape resembles that of *Reganochromis calliurus* (see page 257), has been collected, as far as is known, only from two locations in the south, Hore Bay and Sumbu Bay (Bailey & Stewart, 1977); no record of any additional collection could be found. The head is very elongate and resembles that of a pike. Even more striking is the typical dorsal fin: the membranes between the first five spines are elongated and form pointed lappets like those found in several *Apistogramma* species from South America. Unfortunately this species has been caught only at great depths, between 40 and 100 metres, so no information on its ecology is available.

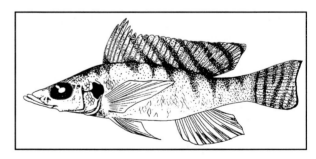

Baileychromis centropomoides (after Bailey & Stewart, 1977).

Piscivores of the depths

It has been known for a long time that the species belonging to the genus *Bathybates* have very characteristically coloured males which are very attractive. Apart from a few odd specimens, usually of *Bathybates minor*, these predators are rarely collected for the aquarium trade, but, because of their excellent taste, they are commonly fished for using hook and line.

Very little is known of the biology of these fishes, a prerequisite for finding an area where they might perhaps be caught in shallow water (in order to collect them alive); it is assumed that some of them do, at least sometimes, venture into the upper water layers because they are regularly seen in the catches of the Capenta or Ndaga fishermen, who fish primarily for the small lake sardines in surface waters. Of the few specimens exported to date, almost all were *B. minor* and juvenile *B. fasciatus*, the two species apparently most often seen among the schools of sardines in surface waters; both species are also regularly found at depths of between 120 and 200 metres (Coulter, 1991).

B. minor, as its name indicates, is the smallest *Bathybates* species. It lives among, and preys on, the huge shoals of the lake sardines, *Stolothrissa tanganicae* and *Limnothrissa miodon*, behaving not as a pursuit hunter, but instead making surprise attacks on its victims. Its coloration and size are similar to those of its prey, and it seems likely that the sardines mistake the voracious predator for one of their own kind. One of the photos (page 137) shows that *B. minor*, although rather small, feeds on prey of a size similar to itself. In the aquarium its feeding behaviour remains typical. The food is targeted and seized with a lightning dart, and large chunks are swallowed whole. The size of the food morsels that it can fit into its mouth is remarkable.

The maximum total length of *B. minor* is about 20 cm, but most of the specimens I have seen (around the lake and in importers' tanks) measured between 8 and 12 cm. All *Bathybates* are mouthbrooders and have very large eggs. Poll (1956) reports a 17.5 cm long brooding female which held 60 large (approx. 6 mm) eggs! *B. minor* has spawned in the aquarium (Allen, 1996) but details of its spawning behaviour are not yet known. Allen found that the female is larger than the male and that males do not prepare a spawning site. The male colour pattern of four horizontal black stripes is seen only at breeding time. Females and sub-adult males are very silvery without any markings.

Bathybates fasciatus and *B. leo* are two further common species which appear to feed exclusively on lake sardines and are usually found together with their prey. Interestingly Poll (1956) reports that *B. fasciatus* could be caught at night with hook and line using sardines as bait. Juvenile *B. fasciatus* seem to mix with the sardines and are frequently collected with them in surface waters. In the aquarium they display a behaviour different from that of *B. minor*. *B. fasciatus* is a true pursuit hunter and needs a very large tank in which it can hunt its prey. I have been able to keep *B. fasciatus* on a diet of regular aquarium fare, even flake food (!), but the single specimen I kept met its death by getting stuck under a rock while hunting a small fish that hid there. *B. fasciatus* normally hunts in open water where it swims swiftly after its prey; obstacles such as rocks are hardly ever found in its path.

B. leo and adult *B. fasciatus* are usually found in deep to very deep water, and both have been caught at depths of 200 metres, just above the anoxic zone of the lake (Poll, 1956). Adult specimens are normally caught by hook and line. *B. fasciatus* is the only species of the seven known that I have seen underwater. On three occasions two or three individuals were seen a few metres above the bottom at a depth of 35-40 metres. On another occasion I have seen a very large individual with male coloration in water as shallow as 10 metres (Kala Island, June 1994). On all these occasions I was

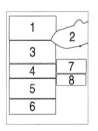

page 136

1 and 2. *Bathybates minor* in the aquarium.
3. *Bathybates graueri* in the aquarium.
4. A female *Bathybates leo* at Chimba, Zambia.
5. *Chrysichthys graueri* (Chituta).
6. *Chrysichthys stappersii* (Chituta).
7. The two most common species of herring: *Limnothrissa miodon* (top) and *Stolothrissa tanganicae* (bottom); (Bulu Point, Tanzania).
8. *Plecodus elaviae*.

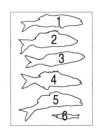

page 137

1. *Bathybates fasciatus*.
2. *Bathybates ferox*.
3. *Bathybates horni*.
4. *Bathybates leo*.
5. *Bathybates vittatus*.
6. *Bathybates minor* with prey still in its mouth.

able to catch only a glimpse of the fishes as they maintained a large flight distance.

The maximum recorded size of *B. fasciatus* is 41 cm and that of *B. leo* 35.5 cm (Coulter, 1991), and males of both species have a very attractive pigmentation pattern. An adult male would, of course, be a spectacular inhabitant for any large aquarium, but it has yet to be determined whether or not these pursuit hunters can adequately be kept in captivity. As in all *Bathybates*, females are silvery and exhibit a faint pattern paralleling the markings found in males. It seems that *B. leo* and adult *B. fasciatus*, together with *B. minor,* are the only species of this genus which are found in the open water column, with the first two normally found at great depths. It is thus possible that breeding in these three species may take place in open water. Juvenile *B. fasciatus* are regularly caught in shallow water over sand, perhaps indicating that females release their offspring in such habitats.

Juvenile *B. ferox* can sometimes be found in beach seine catches together with young *B. fasciatus* and *B. minor*. *B. ferox* is another large species with a recorded maximum total length of 36 cm, but it is not as streamlined as the pursuit hunters *B. fasciatus* and *B. leo*. Stomach contents analysis of *B. ferox* has revealed mainly bottom-dwelling cichlids (e.g. *Xenotilapia* spp.), so it apparently exploits a different food source to the sardine-hunters of the open water. Adults have also been found in shallow water, although it is not known whether they came there to hunt or to breed, or were females releasing their offspring. Poll (1956) recorded a gravid female with a swollen abdomen completely filled with 6.5 mm large eggs. The greatest depth at which *B. ferox* has been caught is 70 metres, whereas all the other species, with the exception of the rare *B. horni*, have been caught at depths of more than 160 metres.

B. horni seems to be a rare species (Coulter, 1991) and has hardly ever been caught by hook and line. I have seen some specimens which were caught in a gill net near Kipili together with *B. ferox*. *B. horni* is streamlined like *B. leo* and *B. fasciatus* and may thus be a pursuit hunter, albeit perhaps in a different area or habitat. It can attain a total length of more than 30 cm.

B. graueri also attains a maximum length of about 30 cm and adult specimens are sometimes caught in shallow water. *B. graueri* is normally found near the bottom where it preys on sand-dwelling cichlids such as *Xenotilapia* spp. (Coulter,

1991). In the aquarium, however, it behaves rather like an open-water species, rarely swimming close to the bottom. It thus seems likely that *B. graueri* cruises at a certain distance from the bottom while searching for prey.

The seventh species, *B. vittatus*, also preys on cichlids rather than sardines. Some individuals have been taken at depths of more than 200 metres and others in shallow areas over a muddy bottom. Maximum recorded length is 42 cm, which makes it the largest member of the genus. It has a remarkable scale pattern in which "normal"-sized scales are embedded among numerous tiny scales, but the function of this type of squamation, as with many other features of *Bathybates*, is unknown.

It is likely that the breeding behaviour of *Bathybates* species will be determined only by dedicated aquarists, as probably all species breed at a deep level, too deep for direct observation by divers using SCUBA gear.

Hemibates stenosoma has a colour pattern similar to that of *Bathybates* but has a deeper body and much smaller teeth. Maximum total length is 27 cm, which makes this fish a respectable predator. *H. stenosoma* has a lake-wide distribution and is sometimes caught in large numbers in gill nets. Like *Trematocara* species it migrates to shallow water during the night, and, as it is often found together with members of that genus, it quite probably also preys on these little invertebrate-feeders.

H. stenosoma is a maternal mouthbrooder which produces some of the largest cichlid eggs known: 7 mm (Poll, 1956). Males have a beautiful colour pattern of black blotches and streaks whereas the females are completely silvery in colour. Interestingly there seems to be quite some variation in the male colour pattern. Some individuals collected in Chituta Bay, Zambia, had distinct vertical bars on the anterior part of the flank whereas other males in the same catch exhibited irregular blotches. The ones with the vertical bars seemed to be larger; the local fishermen call both variants "Mphandi" and also know that the females are the plain-coloured ones. The holotype has a pattern of irregularly-shaped blotches.

It is possible that the male colour pattern changes with age, but equally this variation could be analogous to the subtle differences in the colour patterns of the seven *Bathybates* species, and we could thus be dealing with two different *Hemibates* species.

The open water

The open water habitat — the water column above the rocky substrate — is not really a well-defined habitat and could easily have been included among those previously discussed. However, the cichlids we find here — the plankton-feeders and their predators — have traditionally been treated separately. Interestingly several species of this group have breeding territories in the open water as well, although they still seem to be dependent on the rocky substrate.

Microdontochromis

Microdontochromis species are normally found in the upper five metres of the open water, and although they occur in rocky habitats they are more often seen over intermediate habitats where the bottom is mainly sandy with some rocks. There are two species known, the main difference between them being the horizontal mid-lateral line of black spots present in *M. tenuidentatus*. There are some morphological differences as well: *M. rotundiventralis* has two to three rows of teeth while *M. tenuidentatus* has a single row, and, as the name suggests, the ventral fins of *M. rotundiventralis* are rounded whereas the last rays of the ventral fins in *M. tenuidentatus* are longer than the first.

The holotype of *M. tenuidentatus*, together with about 50 other specimens, was collected in the Congo near Livua (just north of Moliro). The holotype of *M. rotundiventralis* was caught at Nkumbula Island (Mbete Island) near Mpulungu. I have found *Microdontochromis* at Chimba and Cape Kachese in Zambia; these lacked the mid-lateral band of black spots and had rounded ventral fins, and can thus be classified as *M. rotundiventralis*. Chimba, however, is rather close to Livua (the type locality of the other species), and because the two species have not yet been found sympatrically, the only barrier in their dispersal could be Livua Bay or Moliro Bay. It is hard to imagine that two species with such a similar morphology and close relationship could be found sympatrically, which would indeed prove that we are dealing with two different species and not two geographical variants of a single taxon.

Microdontochromis have also been seen at Cape Tembwe, Congo, where they exhibit the horizontal row of black spots on the body, which is also common in specimens from Burundi. We can conclude from these data that *M. tenuidentatus* is found in Burundi and almost all of the Congo, and its range extends into Tanzania at least as far as Miyako, near Bulu Point (Kuwamura, 1986a). I have seen *M. rotundiventralis* at Kala Island in Tanzania, so its range probably encompasses all of Zambia and part of the southeastern shore in Tanzania.

Microdontochromis live in shallow water, in large schools of sometimes more than a thousand individuals. They feed on zooplankton, picking out individual organisms with their rather protractible mouths. The head is laterally very compressed, which appears to be an important characteristic of these species. The teeth in the outer (maxillary) jaws are minute and look rather fragile. The lower pharyngeal bone is very thin and

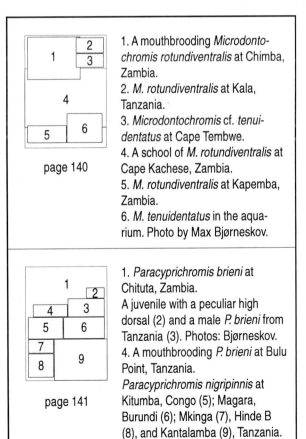

1. A mouthbrooding *Microdontochromis rotundiventralis* at Chimba, Zambia.
2. *M. rotundiventralis* at Kala, Tanzania.
3. *Microdontochromis* cf. *tenuidentatus* at Cape Tembwe.
4. A school of *M. rotundiventralis* at Cape Kachese, Zambia.
5. *M. rotundiventralis* at Kapemba, Zambia.
6. *M. tenuidentatus* in the aquarium. Photo by Max Bjørneskov.

page 140

1. *Paracyprichromis brieni* at Chituta, Zambia.
A juvenile with a peculiar high dorsal (2) and a male *P. brieni* from Tanzania (3). Photos: Bjørneskov.
4. A mouthbrooding *P. brieni* at Bulu Point, Tanzania.
Paracyprichromis nigripinnis at Kitumba, Congo (5); Magara, Burundi (6); Mkinga (7), Hinde B (8), and Kantalamba (9), Tanzania.

page 141

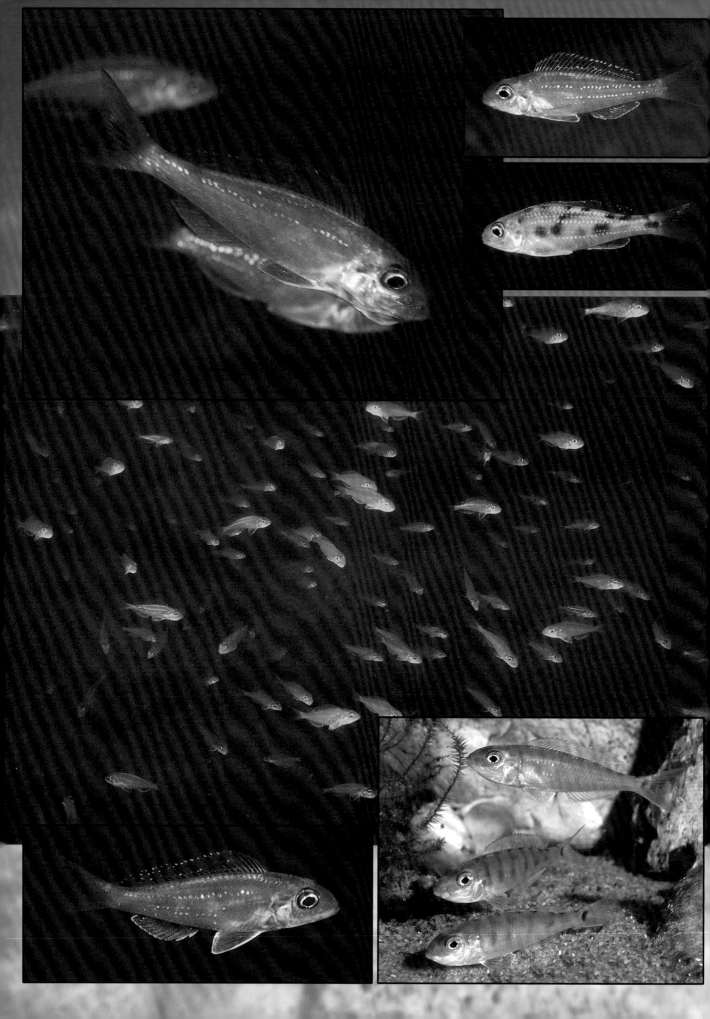

slender, and the teeth on this bone are minute and pointed, a common feature of plankton-feeding cichlids.

M. tenuidentatus and *M. rotundiventralis* are mouthbrooders. The first reports of the breeding biology of the latter (Puttberg in Konings, 1991b) suggested that it was a maternal mouthbrooder. Puttberg noted that the eggs are rather large and that the female broods only a few larvae to maturity, although about 20 eggs are deposited during spawning. The male did not seem to make a nest or defend a territory. The eggs were brooded for about four weeks before the fry were released. During this period the female continued feeding as if she had nothing in her mouth *not* to be swallowed.

More recently, however, Yanagisawa *et al.* (1996) found that fry with a length of less than 6 mm were brooded by the female but that larger fry were also taken up by the male; we must thus conclude that under natural circumstances *M. rotundiventralis* is a biparental mouthbrooder. Both mouthbrooding parents feed normally, and fry inside the buccal cavity feed on anything entering the mouth, showing a tenfold increase in weight during the incubation period.

Paracyprichromis

Paracyprichromis contains two species, *P. nigripinnis* and *P. brieni,* which were previously included in the genus *Cyprichromis,* but were placed in the new genus *Paracyprichromis* by Poll (1986) because of differences in the number and arrangement of the vertebrae. Although this splitting may seem questionable, *P. nigripinnis* and *P. brieni* are probably not closely related to the species currently in *Cyprichromis,* and in my opinion this is simply a remarkable case of parallel evolution in which taxa of different ancestry have developed into similar looking species. The most important distinction between the two genera, however, lies in their spawning behaviour. The eggs of *Cyprichromis* are fertilized inside the female's mouth and those of *Paracyprichromis* are not. Spawning in *Paracyprichromis* takes place alongside the vertical face of a large rock, almost in immediate contact with the substrate. In most *Cyprichromis* species spawning takes place in mid-water. During spawning female *Paracyprichromis* swim head-down, close to the substrate, and discharge the eggs. The eggs are collected by the female as they pass by her head; meanwhile the male stays above or beside her and discharges his milt continuously — clouds of seminal fluid are occasionally seen. The male fans the milt towards the falling eggs which are thus fertilized before the female picks them up.

Paracyprichromis nigripinnis and *P. brieni* are both plankton feeders although they maintain close contact with the rocky substrate. Male *P. nigripinnis* are often found in caves or close to the underside of rocky ledges, and in dark places such as these they are usually found swimming upside-down, with their backs towards the light reflected from the bottom. Female and non-territorial *P. nigripinnis* feed from the plankton in the water column one to two metres away from the substrate.

P. nigripinnis occurs mostly at depths below 25 metres, where it is common in rocky caves. At Hinde B in Tanzania I have seen *P. nigripinnis,* or the "Blue Neon" as it is known in the hobby, as shallow as 6 metres, but this is exceptional. *P. brieni* lives in the rocky habitat as well, but males defend territories on the outside of the rocky substrate rather than in caves. This species may be found in shallow water but usually spans a large range from the shallows down to depths of 30 to 40 metres. Non-territorial individuals feed in the open water column, but they too never venture far from the rocks.

Male *P. nigripinnis* defend a territory inside a cave where spawning sites are commonly found next to the ceiling or alongside the upper vertical walls. Females are not attracted to the nest by fin display, as is common in lek-breeding cichlids, but are sequestered from the school and then led to the male's territory.

By contrast the territory of *P. brieni* is on the outside surface of a large boulder, most often alongside its vertical face.

P. nigripinnis has a lake-wide distribution and all populations known have similarly coloured individuals. The "blue neon" stripes are prominent in males but less spectacular in females. In the dark of the male's cave the iridescent stripes on his fins and body beautifully indicate his position to entering females.

The holotype of *Paracyprichromis brieni* was collected along the Ubwari Peninsula in the Congo, but this species too enjoys a lake-wide distribution. Geographical variation is very apparent and some variants have even been classified as different species by aquarists; some variants from various localities around the lake are shown on page 144. *P. brieni* and *P. nigripinnis* are usually found sympatrically and both species can be found living side by side.

The maximum total length of *Paracyprichromis nigripinnis* is about 10 cm. Its shape closely resembles that of *P. brieni* which has a similar maximum size. Anatomically, *P. brieni* can be distinguished from *P. nigripinnis* by a smaller eye in relation to the length of the head. Juvenile *P. brieni* can be recognized by the yellow anal fin. Females of the northern population have yellow anal fins as well; those of the southern populations have a black marginal band in the dorsal and anal fin.

Cyprichromis

The genus *Cyprichromis* currently consists of three described species (*C. leptosoma, C. microlepidotus,* and *C. pavo*) and two undescribed forms (*C.* sp. "leptosoma jumbo" and *C.* sp. "zebra"). Species of the genus *Cyprichromis* forage in schools that can number more than 10,000 individuals. Such schools can consist of more than a single species and it is not uncommon to find three *Cyprichromis* species — *C. leptosoma, C. pavo* (or *C. microlepidotus*), and *C.* sp. "leptosoma jumbo" — foraging from the plankton side by side. Interestingly there are several different types of schools. The foraging schools mostly contain adult females and non-territorial males. Closer to the rocky substrate, schools consisting solely of territorial males appear to be more static, and ripe females ready to spawn are also found in such gatherings. Males remain at a fixed distance from their neighbours which can be any one of the four *Cyprichromis* species. The third type of school consists of mouthbrooding females, and although usually consisting of a single species, mixed schools have been seen as well.

Cyprichromis, like many other maternal mouthbrooding species, are typical lek-breeders, but a remarkable feature in their case is that territories are clustered in three dimensions. It is very impressive to observe thousands of males defending their territories in mid-water. They stake out their territories about 100 cm apart (vertically as well as horizontally) and their neighbours may be different *Cyprichromis* species or conspecific males of a different colour. The males remain constantly in mid-water and relate the boundaries of their premises to the distance of neighbouring males. Females are continuously courted and attracted to the male's territory as soon as they enter the three-dimensional breeding lek.

When a female responds to a male's courtship she enters his territory and spawning is initiated. The male bends his body and all of his fins (except the ventrals) away from the female. The ventral fins, which have a large yellow tip, are held to the outside and vibrated in front of the female (see photo page 145), and she responds by snapping at them several times before egg-laying begins. The snapping may indicate that the male discharges milt before any eggs are laid. Since spawning takes place in mid-water, the ventral fins may indicate to the female the spot where she is supposed to deposit her eggs, i.e. in the centre of the male's territory. The colour of the ventral fin tips does not exactly match that of the eggs and it may instead be the vibration that triggers the female to snap at them. I have also observed males snapping at the displayed fins of a more dominant male.

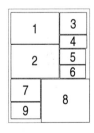

1	7
2	8
3	9
4	10
5	11
6	12

page 144

Paracyprichromis brieni at:
1. Milima Island, Congo.
2. Kitumba, Congo; male.
3. Kitumba; female (aquarium).
4. Kapampa, Congo; male.
5. Cape Kachese, Zambia.
6. Kapemba, Zambia.
7. Magara, Burundi; female.
8. Bulu Point, Tanzania.
9. Msalaba, Tanzania.
10. Nkondwe Island, Tanzania.
11. Isanga, Zambia
12. Chituta, Zambia.

1	3
	4
2	5
	6
7	8
9	

page 145

1. *Cyprichromis pavo* at Samazi, Tanzania.
2. *Cyprichromis* sp. "zebra" at Chituta, Zambia.
3. *C. pavo* at Kantalamba, Tanzania.
4. *C. pavo* at Chituta.
A juvenile *C.* sp. "leptosoma jumbo" (5) and a juvenile *C. leptosoma* (6) at Kambwimba, Tanzania.
7 and 8. Spawning *C.* sp. "leptosoma jumbo" at Kekese, Tanzania.
9. *C. leptosoma* at Kekese.

Next the male positions himself over the female, all fins extended, and nudges her gently on the head with his fully opened mouth (see photo page 145). The female releases an egg (or a few) and immediately swims backward to retrieve it (them). This sequence of release and backing-up may be repeated several times until the female has moved too close to the boundary of the male's territory and has to be led back to its centre. There follows a number of "ventral snappings" before the female discharges the next series of eggs.

As soon as the female ventures too far from the territory's centre neighbouring males have an opportunity to court her and perhaps eventually lead her to their sites. The few times I have been able to observe spawning in the lake, however, it appeared that the females returned to the same male even though neighbouring males were trying to "capture" them. Sneak males (sexually active males without a territory) are common and try to get in on the act by letting the female snap at their vibrating ventrals — usually at the same moment the tenant of the territory "shakes his fins". Sneak males are violently expelled from the territory but are probably successful in fertilizing some of the female's eggs.

After the female has spawned she joins a separate school of mouthbrooding females. After an incubation period of about three weeks the females in the school release their offspring simultaneously. Fry gather in large schools in the upper centimetres of the water column.

Cyprichromis breed according to the typical lek-breeding mechanism seen in many other maternal mouthbrooding species. The remarkable feature is that territories are clustered in three dimensions. Ripe females are normally triggered to oviposition by the sight of a displaying male and the sight of a nest. When no obvious nest (spawning site) is present the male directs the female — see for example *Ophthalmotilapia* (page 42) and *Cyphotilapia* (page 118). A male in mid-water has no reference point that would mark its territory or its spawning site, i.e. the nest is invisible. Therefore the ripe female may be triggered to oviposition by the vibrating pelvics with a concomitant release of sperm by the male. The ingested sperm prior to spawning could be the releasing factor in the female; this could also be true for many other mouthbrooders.

Cyprichromis pavo and *C. microlepidotus*

C. microlepidotus and *C. pavo*, males of which can attain a total length of about 12 cm, are characterized by having small scales on the flank; these are smaller and more numerous than the corresponding scales in other known species of the genus.

C. pavo, the holotype of which was collected at Tembwe in the Congo (Büscher, 1994), has its range in the southern part of the lake. It has been found as far north as Kanoni along the western shores of the lake, while the northernmost location on the east coast is Msalaba (Cape Mpimbwe) in Tanzania. It is common between these two points, but prefers the deeper regions of the rocky habitat and is often found between 25 and 45 metres of depth. Although it has a wide range it does not appear to have formed significantly distinct geographical races.

Its counterpart in the northern half of the lake appears to be *C. microlepidotus*, which has been seen as far south as the Kavala Islands along the west coast and whose southernmost locality is at Bulu Point, Tanzania. The holotype was caught at the Ubwari Peninsula (Poll, 1956). *C. microlepidotus* is known to occur in several geographical races (see photos page 148).

C. microlepidotus and *C. pavo* should probably not be regarded as geographical variants of a single species — even though they are not sympatric — because there are several important differences between the two forms.

Male *C. pavo* have elongated ventral fins and defend rocky sites as a breeding territory. Aquarium observations (Puttberg, in Konings, 1996) have revealed that spawning takes place alongside the vertical face of a rock. This is in contrast to the case in *C. microlepidotus*, which spawns in a manner very similar to the mid-water spawning technique employed by *C. leptosoma*, *C.* sp. "leptosoma jumbo" and *C.* sp. "zebra" (see above).

C. microlepidotus and *C. pavo* males exhibit polychromatism, i.e. differences in colour between males of the same population. In *C. leptosoma*, *C.* sp. "leptosoma jumbo", and *C. microlepidotus* the tail can be either blue or yellow, and other parts of the body may also have variable coloration. Interestingly, in *C. pavo* it is the anal fin, rather than the tail, that can be either yellow or blue, at least in the Zambian and Tanzanian populations. Females of both species have a beige to light-brown colour without markings in the fins.

It seems therefore that *C. microlepidotus* is an intermediate between *C. pavo* and *C.* sp. "leptosoma jumbo". It shares with *C. pavo* the small scales and gross morphology, and with "Leptosoma Jumbo" the variable coloration, geographical variation, and breeding behaviour.

Cyprichromis sp. *"leptosoma jumbo"*

Apart from a small stretch of the Zambian shoreline between Chituta and Kapemba, where it does not occur, the range of *C.* sp. "leptosoma jumbo" is more or less similar to that of *C. pavo*, although it does not seem to have a counterpart in the northern half of the lake. There are said to be two *Cyprichromis* species at the Kavala Islands; one of them is the yellow-coloured *C. microlepidotus* and the other may be *C.* sp. "leptosoma jumbo". I was unable to locate this species at two of the Kavala Islands where I investigated, but some specimens have been exported for the hobby as "Cyprichromis Kibigi". If these specimens were indeed collected at the Kavala Islands, that would extend the range of the "Leptosoma Jumbo" into the northwestern part of the lake. However, "Cyprichromis Kibigi" looks remarkably similar to the Leptosoma Jumbo population found at Tembwe II, which is south of the Lukuga River (the most likely boundary of the species' range).

There are at least eight distinguishable geographical variants of Leptosoma Jumbo known (photos on page 152):

1. Kekese variant (photos 1 and 2). This variant is found between Ikola and Kasoje in Tanzania. The males with a black dorsal do not have a golden yellow head and males with a light-blue dorsal do not have a black spot on the first few spines of the dorsal.

2. Mpimbwe variant (not illustrated). There is to date only a single location known, at Cape Mpimbwe, and this form could also be considered a form intermediate between the Kekese and Malasa variants. Males with a black dorsal also lack the golden-yellow head, but those with a light-blue dorsal have a black spot on the first few spines.

3. Malasa variant (photos 8, 9, and 10). This variant has the widest range of all known geographical forms. Males with a black dorsal have a deep golden-yellow head and those with a light-blue dorsal have a black spot in the anterior part of the dorsal. It occurs between Kipili and the Kalambo River.

4. Isanga variant (photos 11 and 12). This form has black-dorsal males identical to the form found at Kapemba, but at Kapemba I could find no males with a yellow dorsal. At Isanga the males with a light-coloured dorsal have blue tails and yellow dorsal fins, a reversal of the situation found north of the Kalambo River. I have found this variant only along the east side of Chituta Bay.

5. Kapemba variant (photos 13 and 14). This form has, as far as I know, only two colour morphs. Males have a black dorsal fin and either a yellow or a blue tail. The spiny part of the black dorsal has a yellow band at its base and males have a golden-yellow colour on the lower part of the head and shoulder. This variant is only known from Kapemba, Zambia.

6. Chaitika variant (not illustrated). I could find only two morphs: males have a light-blue dorsal, a black anal fin, and either a yellow or blue tail. Although I have seen this variant only at Cape

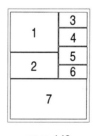

page 148

1. *Cyprichromis microlepidotus* at Bulu Point, Tanzania.
2. *C. microlepidotus* (from Pemba, Congo) in the aquarium. Photo by Volker Puttberg.
3 and 4. *C. microlepidotus* from Magara, Burundi, in the aquarium.
5. *C. microlepidotus* (from Kavala Island, Congo) in the aquarium.
6. Mouthbrooding female *C. microlepidotus* at Bulu Point.
7. *Cyprichromis* sp. "leptosoma jumbo" at Isanga, Zambia.

page 149

Five geographical variants of *Cyprichromis leptosoma*. The yellow-tailed males in the right-hand column, the blue-tailed in the left.
1. and 2. At Kigoma, Tanzania.
3. and 4. At Bulu Point, Tanzania.
5. At Kekese, Tanzania.
6. and 7. At Ulwile Island, Tanzania.
8. At Samazi, Tanzania.
9. At Kantalamba, Tanzania.
10. From Chituta (aquarium).
11. At Isanga, Zambia.

Chaitika it may have a wider range towards the west.

7. Kipimbi variant (not illustrated). This form is known from a few places in Cameron Bay, Zambia, and consists of two different morphs: males have a black dorsal and anal fin and either a yellow or a blue tail. They are very similar to the black dorsal males of the Malasa variant but have less yellow on the head. The Kipimbi variant was the first "Leptosoma Jumbo" exported for the aquarium hobby.

8. Moliro variant (photo 7). I was unable to find males with black fins in Moliro Bay, but there were males with a light-blue to yellowish coloured dorsal fin. Only two morphs were found: blue-tailed and yellow-tailed males. Very similar forms were seen at Cape Tembwe, much further north along the Congolese shore.

9. Kapampa variant (photo 5). Males have a yellowish-blue dorsal with either a yellow or a blue tail. They are further characterized by an orange-yellow colour on the lower part of the head and body. The range of this variant extends between Lunangwa and Tembwe (Tembwe Deux).

10. Kitumba variant (photos 3 and 4). This is possibly the most beautiful variant found to date. Fully adult males are either entirely yellow or blue. Males with a mixture of these two colours are very common (see photo page 153) and this phenomenon is further explained later. This seems to be the most localized geographical variant of the "Leptosoma Jumbo": it was has been found only near Kitumba.

Females at some localities (e.g. Kekese) have a black marginal band in the dorsal and anal fins, but have a deeper body than *P. brieni* females which also can have "black-trimmed" fins.

In almost all populations along the eastern shore there are three different "colour morphs". Apart from the yellow and blue-tailed morphs there are also males with either a black or a light-blue to yellowish dorsal fin. The males with a light-coloured dorsal fin have a yellowish tail. In the wild I have not seen males with a light-coloured dorsal and a blue tail, but in the aquarium such specimens are common. This could suggest that such individuals are the result of hybridization and that each of the three "colour morphs" is in fact a different species! It seems not unlikely that females have a definite preference for a male with a particular colour pattern. In any case these *Cyprichromis* forms have perhaps the greatest po-tential of all known cichlids for the study of the mechanism by which sympatric speciation could perhaps occur (if it ever does!).

Cyprichromis sp. "zebra"

The fourth species, C. sp. "zebra" (also known as the "Wimple Leptosoma") can be recognized by its yellowish coloration, lack of conspicuous colours, and by broad vertical bars which are not permanently exhibited. C. sp. "zebra" does not exhibit polychromatism and has a very restricted range along the western rocky shore of Chituta Bay in Zambia. It is found sympatric with C. pavo and C. leptosoma, but I have not been able to find C. sp. "leptosoma jumbo" near Chituta although it is common along the opposite side of Chituta Bay. C. sp. "zebra" might therefore be regarded as a geographical variant of the Leptosoma Jumbo, but it can easily be told apart from that species.

Cyprichromis leptosoma

C. leptosoma is a small and slender-built cichlid found at virtually all rocky shores with deep and clean water between Kigoma in Tanzania and Mpulungu in Zambia. Apart from the blue colour on the head, it further differs from C. sp. "leptosoma jumbo" in its shallower body and smaller adult size. C. sp. "leptosoma jumbo" can grow to a total length, at least under aquarium conditions, of approximately 12 cm whereas C. leptosoma rarely exceeds a length of 9 cm. C. leptosoma occurs in several geographical variants, some of which can be seen on page 149. Females at most locations have yellowish dorsal fins while those of juveniles are bright yellow.

There are five different geographical variants of C. leptosoma known, and each occurs in both blue- and yellow-tailed morphs. The holotype stems from the Mpulungu area. The variant at Mpulungu is characterized by a light-blue dorsal fin speckled with small black dots in the blue-tailed morph, and by a black band at the base of the blue dorsal in the yellow-tailed morph. This variant is found at Mpulungu, Chituta, Isanga, and across the Kalambo River outlet up to Kasanga in Tanzania (photos 9, 10, and 11 on page 149).

North of Kasanga, at Samazi, we find the so-called "Malasa Leptosoma". This variant has a blue dorsal fin and males with a yellow tail have an additional yellow spot in the soft-rayed part of

150

the dorsal. Blue-tailed males have a bluish dorsal and sometimes a darker spot is visible in the soft-rayed part of the fin. They resemble those from Bulu Point, but the dark spot in the dorsal is not as prominent. The Malasa variant occurs between Samazi and Kipili, and probably as far as Cape Mpimbwe (photos 6, 7, and 8).

The third variant is known from the rocky coast north of Ikola, but it may have a wider range extending along the coast north of Sibwesa, up to Kasoje. The yellow-tailed males are characterized by a bright yellow anal fin and the blue-tailed males have a yellow dorsal fin (photo 5). The fourth variant, the "Karilani Leptosoma", occurs north of Kasoje, around Karilani Island and at Bulu Point. It resembles the "Malasa" variant, but the yellow-tailed males do not have a yellow spot in the dorsal. The blue-tailed males, however, have a dark spot in the soft-rayed part of the dorsal and an orange-yellow body (photos 3 and 4). The fifth variant is found around Kigoma and is characterized by a yellowish head. It does not have spots in the dorsal fin (photos 1 and 2).

Mixed-species schools

Schools consisting of males of three different *Cyprichromis* species appear to contain many more species, as each of the three is represented by at least two different colour morphs. I have suggested previously (Konings, 1988) that the blue-tailed individuals of, for example, *C. microlepidotus* could in theory be a species genetically different from the yellow-tailed specimens. Their behavioural preference for feeding from plankton above rocky substrates may have brought these two hypothetical species together, just as it brought *Paracyprichromis*, which has a similar feeding behaviour but is from a different ancestry, together with *Cyprichromis*.

The observation that a female may give birth to yellow as well as blue-tailed males in captivity does not prove the existence of polymorphism in nature. Such a female could, for example, belong to a yellow-tailed "species", but have spawned with a blue-tailed male in the confines of the aquarium. If the colour of the tail constitutes the main criterion for mate recognition, she would never have done so in the wild. Until unambiguous experiments have been performed to determine the inheritance of the colour of the tail, the possibility of the existence of two or more sympatric species cannot be excluded.

Sometimes, when such "hybridization experiments" seem to have occurred in the wild, we may be led to draw false conclusions regarding the specific status of the different sympatric species. This may be the case with the population of *Cyprichromis* sp. "leptosoma jumbo" at Kitumba, Congo. In this population not only the colour of the tail differs, but also that of the body. In fact there are three different morphs: there are completely yellow males and blue males with yellow or blue tails. I was able to observe this school for only an hour, but even during this short period I gained the impression that the completely yellow individuals and the blue individuals represented two different species. As is usual among *Cyprichromis*, males defend their three-dimensional territory in the open water and relate the boundaries

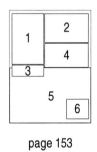

1	2
3	4
5	6
7	8
9	10
11	12
13	14

page 152

Geographical variants of *Cyprichromis* sp. "leptosoma jumbo".
1 and 2. Kekese, Tanzania.
3 (photo by Volker Puttberg) and 4. Kitumba, Congo.
5. At Tembwe II, Congo.
6. "Cyprichromis Kibigi" (from Kavala?) in the aquarium.
7. Moliro, Congo.
8. Kantalamba, Tanzania.
9 and 10. Kambwimba, Tanzania.
11 and 12. Isanga, Zambia.
13 (aquarium) and 14. Kapemba, Zambia.

	2
1	4
3	
5	
	6

page 153

1. The change in colour pattern of the same male *C.* sp. "leptosoma jumbo" from Kitumba, Congo, over a period of 18 months (the third image shows the fish from the other side). Photos by Volker Puttberg.
2. A school *Cyprichromis* spp. at Fulwe Rocks, Tanzania.
3. *Tangachromis dhanisi*.
4. *Haplotaxodon microlepis* at Kisambala, Tanzania.
5. A brood of *H. microlepis* fry seek shelter in the male's mouth (Kisambala).
6. A female *H. microlepis* and her offspring at Ulwile Island, Tanzania.

to the distance from neighbouring males. Yellow and blue males were found mixed, and among them large numbers of females, which all looked identical, were trying to find spawning partners. Both types of male sometimes courted the same female, but their territorial aggression was always directed towards a males of the same colour; i.e. yellow males would chase only other yellow males while blue males (with blue or yellow tails) chased only other blue males. Sometimes a chase followed a route through the territory of a differently coloured male!

After a while I noticed a few rare individuals which looked like a cross between the two morphs. The colour aberrations I noticed were mainly of the nature of a gradual change from blue to yellow rather than a combination of the (putative) parental colours. There was a mainly yellow individual with a blue blotch on its body or *vice versa*, i.e. a blue male with a yellow spot.

Puttberg (1996) found that sub-adult individuals that show yellow pigment at an early stage of their development eventually turn into completely yellow specimens. Although it cannot be denied that the yellow males are distinctly different from the blue ones (the ones with a blue tail), it is still not clear whether females appreciate these differences as well.

These observations can be explained in several different ways, of which the presence of two sympatric but distinct species is just one. The yellow and blue males could represent different morphs of a single species with the females spawning with all colour morphs. The territorial colour of the males could be regulated by one or very few genes which could act like a switch in a juvenile phase of the male: normally it would switch on either blue (tail) or yellow (tail). This could occur as a phenotypic event, independently from genetic factors.

If the males of this and other populations are to be regarded as morphs of one species then they certainly present a remarkable case of polymorphism. If such contrastingly coloured males can belong to one species then how can we explain the existence of three species of *Cyprichromis* at most places in the southern half of the lake? The colour difference between the males of these three species is much less dramatic than that found in the Kitumba population of the "Leptosoma Jumbo". Maybe the latter population can teach us a lesson as regards speciation. It may turn out that colora-

tion is not the most important criterion in species recognition but that species specific odours combined with behavioural and morphological characters play a major role instead.

In *C. microlepidotus* there are blue-tailed males with sky-blue heads and yellow tailed males with variable black blotches in the dorsal and on the back. In the aquarium, males of one morph seem to regard those of the other as conspecific and chase them out of their territories. Because of the (assumed) importance of male coloration in mouthbrooding cichlids it would be interesting to know if females, in this species too, have a preference for one of the two males. If so, this would mean that here too there may be two species continuously mingled with each other. Under natural circumstances the female would unerringly spawn with the right male since the differences in colour are prominent. Such a scenario is rather commonplace in some other African lakes; for example in Lake Malawi and Lake Victoria many species are distinguished solely by virtue of the male's coloration.

A possible advantage for *Cyprichromis* having differently coloured males which do not visually recognize each other as belonging to the same species, is that their territories can be closer together without eliciting constant territorial fights between neighbouring males of similar appearance. Indeed, it is not uncommon to find yellow-tailed and blue-tailed territorial males almost perfectly alternating — all three species combined — in the water column.

Haplotaxodon microlepis

Haplotaxodon microlepis is a predator which is usually encountered in the upper 10 metres of the water column. It is specialized in preying on small fishes living in this upper layer. Juvenile fishes of some species, such as *Lamprichthys tanganicanus*, *Limnothrissa miodon*, *Ophthalmotilapia ventralis*, and *Cyathopharynx furcifer*, form large schools in the upper centimetres of the water column. Although several species are found in such schools their members all have a similar size (and thus similar swimming speeds). *H. microlepis* can frequently be observed feeding on such schools from below, its almost vertical, upward-pointing, mouth allowing it to stalk its prey from this station. Pursuit hunters and other piscivores cannot easily pursue these small fishes in the upper centimetres of the water column, so *H. microlepis* may have evolved specifically to occupy this niche. At times when young fishes are scarce *H. microlepis* feeds on other types of food, such as zooplankton. *H. microlepis* has a lake-wide distribution and no geographical variants are known.

The maximum total length of this predator is about 26 cm, but average total length varies around 19 cm for males and 17 cm for females. Sub-adult *H. microlepis*, which seem to feed more on plankton than small fishes, are commonly found in schools numbering hundreds of individuals. Upon reaching maturity pairs separate from the school and breed. Breeding in this species appears to occur throughout the year. *H. microlepis* is a biparental mouthbrooder, and male and female stay together during the brooding period, which can be as long as two months. The eggs and larvae are brooded by the female only (Kuwamura, 1988) and when the fry have attained a length of about 9 mm (10 to 12 days after spawning) they are released for the first time. Subsequent mouthbrooding of the fry is mainly the male's task, but the female sometimes helps as well, in particular when the fry get larger. The fry are most closely guarded by the female while the male scouts the surroundings of the release area, but when danger threatens they find shelter inside the male's mouth. Spawns usually number between 100 and 200 eggs (Kuwamura, 1988).

A scale-eater

In terms of total numbers, *Plecodus paradoxus* is possibly the most common scale-eater to be found in Lake Tanganyika. It sometimes occurs in large schools in the northern part of the lake (Brichard, 1978) and has been caught at depths as great as 250 metres!

Plecodus paradoxus is rather common over sandy habitats, often near rocks. It is usually found in groups, rarely on its own. Foraging groups normally swim in a loose formation about one metre above the substrate. Although there are several other silvery-coloured, elongate, scale-eaters to be found in the lake, *P. paradoxus* can easily be distinguished from these by the black spot on the caudal peduncle. When observed in their natural habitat, non-breeding *P. paradoxus* exhibit a narrow mid-lateral stripe which runs from the edge of the gill cover to the spot on the peduncle. Adults can reach a total length of about 25 cm.

The main point of interest is not its peculiar feeding habit but the fact that *Plecodus*, together with *Perissodus*, exhibits breeding behaviour intermediate between maternal mouthbrooding and substrate brooding. I have observed brooding and fry-guarding parents of *Per. microlepis*, *Per. eccentricus*, *Pl. straeleni*, and *Pl. paradoxus* and found that their breeding behaviour is apparently very similar. Yanagisawa & Nshombo (1983) made a very detailed study of *P. microlepis*, and to my mind the other species behave in the same or a very similar way.

The breeding behaviour of these scale-eaters can be described as follows: spawning takes place on a hard substrate; the female picks up all the eggs, which are very small and numerous; after about 9 days the fully-developed but tiny fry are released from the female's mouth; a few days later the fry increasingly seek refuge inside the male's mouth instead of in that of the female; both parents guard their offspring for at least six weeks.

I have several times been able to observe how frightened fry of *P. microlepis* were taken up by their parents. This process takes only a few seconds and the fry are then safe inside the parents' mouths. I also observed two fry-guarding *P. paradoxus* pairs which had both chosen to make their nest on top of a large boulder at a depth of

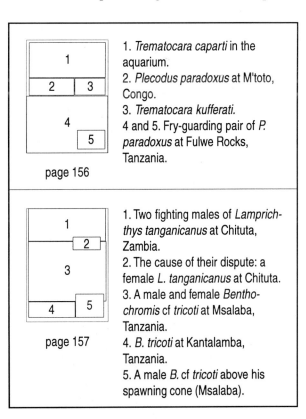

page 156

1. *Trematocara caparti* in the aquarium.
2. *Plecodus paradoxus* at M'toto, Congo.
3. *Trematocara kufferati*.
4 and 5. Fry-guarding pair of *P. paradoxus* at Fulwe Rocks, Tanzania.

page 157

1. Two fighting males of *Lamprichthys tanganicanus* at Chituta, Zambia.
2. The cause of their dispute: a female *L. tanganicanus* at Chituta.
3. A male and female *Benthochromis* cf *tricoti* at Msalaba, Tanzania.
4. *B. tricoti* at Kantalamba, Tanzania.
5. A male *B.* cf *tricoti* above his spawning cone (Msalaba).

about 15 metres (Fulwe Rocks, Tanzania). When I disturbed these pairs they started to collect their fry by holding their mouths open among the school. It took each pair more than two minutes before all of the still very small fry were safe inside their mouths. After about 15 minutes the fry were released at the same site. The two broods contained about 200 fry each, which seems to be a common size among scale-eaters.

The princess of the deep

One of the most spectacular cichlids found in the lake is *Benthochromis tricoti*. It is a very graceful cichlid which occurs at depths of 25 metres or more. Non-breeding individuals gather in large foraging schools in the very deep water (deeper than 50 metres) and feed from the plankton. The maximum total length of a male *B. tricoti* is about 18 cm, females remain a few centimetres smaller.

Poll (1948) noticed that in the collection of *B. tricoti* there were specimens with a rather different colour, but morphologically they seemed identical to the holotype of *B. tricoti*. Later (Poll, 1984) he revised his ideas and described the differently coloured specimens as *B. melanoides*. The holotype of *B. tricoti* was collected near Moba and the holotype of *B. melanoides* near Kalemie, both localities in the Congo. *B. melanoides* is characterized by a large dark blotch on the nape of the head and by the lack of dark horizontal stripes on the body.

Almost all *B. tricoti* available in the hobby originate from the Zambian part of the lake and many of them were caught in Chituta Bay. Either *B. tricoti* has developed into geographical variants or there are at least three different species of *Benthochromis* known to occur in the lake. Although *B. melanoides* could be regarded as a geographical variant (nothing is known about its behaviour and its range), the form I found at Msalaba (Cape Mpimbwe) differs not only in coloration but also in breeding behaviour. The *B. tricoti* in Zambian waters resembles the holotype in its coloration, and I have seen similarly coloured individuals in Congolese as well as in Tanzanian waters.

Males defend a breeding territory which consists of a spawning site atop a large boulder, usually at a depth of 25 metres or deeper. Such boulders stand out from the surrounding rocks. The distance between individual territories is usually more than 10 metres. Males at Msalaba, however, had constructed sand turrets on top of which spawning took place (see photo page 157). In the same area there were sufficient large boulders present but several males had chosen to build their nest themselves. The nests were only about 5 metres apart and females went around spawning with the various males. Such behaviour is rare in Lake Tanganyika — only *Callochromis* species build similar sand nests.

In the aquarium *B. tricoti* has been known to spawn on horizontally-placed flat slabs, along the vertical side of the aquarium, and also on top of sand that has been heaped in the corner of the tank (Salvagiani, 1996). It could be that *B. tricoti* normally builds sand cone nests but on steep rocky coasts adopts the tops of large boulders as spawning sites.

Apart from the different spawning sites, the form at Msalaba also differs in coloration and the shape of the dorsal fin. Males lack most of the horizontal stripes on the body, and the soft-rayed part of the dorsal fin is rounded whereas it is pointed in *B. tricoti* males in the southern part of the lake. The Msalaba *Benthochromis* could be a geographical variant of *B. melanoides*, which also lacks the horizontal stripes, but more information about that form needs to be available in order to establish that it is indeed a different species and not a geographical variant of *B. tricoti*.

Breeding has been observed in the aquarium (Salvagiani, 1996). It seems that the female inspects and cleans the spawning site of sand grains and other irregularities before she lays eggs. The male leads the female to the centre of the nest and leaves the nest while the female starts laying eggs. The female drags her clearly visible ovipositor over the stone and lays a single egg. As soon as the egg is laid she swims backwards, raises the rear part of her body, and picks up the egg. Swimming backwards is also seen in *Cyphotilapia frontosa* and in *Cyprichromis* species, whereas most maternal mouthbrooders swim in circles to collect just-laid eggs. The eggs have a diameter of about 2 mm and a maximum of about a dozen are laid (Salvagiani, 1996). While the female lays eggs the male swims in tight circles above the female with his mouth wide open and all fins erect. Only now and then does he descend to the nest and lie across the female's path. When his genital region comes close to the female's mouth the male quivers and discharges some milt, sometimes visible as a milky cloud, which is then drawn into the female's mouth by opening and closing the gill covers several times.

The length of the mouthbrooding period in *B. tricoti* is not yet known, but the larvae absorb their yolk sacks within the first 10 days of incubation. At this time they are about 8 mm long and barely able to swim (Salvagiani, 1996). It is thus likely that the few fry stay in the female's mouth for a much longer period and feed inside her mouth. Females with a few 2-3 cm long fry have been caught (Toby Veall, pers. comm.) and this could also be the reason that only a few of the tiny eggs are laid: there wouldn't be enough room in the female's mouth for many such large fry. Perhaps the predation pressure on tiny *Benthochromis* fry has led to the development of such a breeding mechanism.

More plankton-feeders from the deep

Some zooplankton-feeders are found at depths of 80 metres or much more. They may live in the same schools as the herring-like "lake sardines", *Stolothrissa* and *Limnothrissa*, which are caught mostly at night when they are attracted to the fishermen's lights. By contrast, the plankton-feeders, *Trematocara caparti*, *T. kufferati*, and *Tangachromis dhanisi*, are only rarely caught on such occasions,

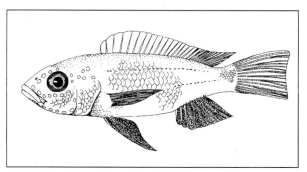

Trematochromis schreyeni (after Poll, 1987).

and they may thus stay at great depths at night as well. The maximum total length of the two *Trematocara* species is about 7 cm. Like other members of the genus they probably are mouthbrooders although very little is known about these deep-living cichlids.

T. dhanisi has a maximum recorded length of 8.5 cm and is probably a mouthbrooder. The male has a yellow-green colour with a conspicuous spot on the operculum. The unpaired fins have a thin black marginal band. The ventral fins are elon-

gated and extend to the caudal peduncle. Another characteristic is the large eye, which is probably a function of the dim light present in the natural habitat. There is no extension of the sensory system such as is seen in the *Trematocara* species.

Poll (1987) has described a species that has close affinities with *Trematocara*: *Trematochromis schreyeni*. The general appearance of this small cichlid (about 7 cm total length) resembles that of the *Cyprichromis*-like species. The mouth is very protractible, which favours the idea that *T. schreyeni* is a plankton-feeder. The most remarkable feature is the enlarged cephalic pores which cover the head almost completely. In contrast to *Trematocara* this species has two fully developed sections of the lateral line — these are absent or only rudimentary in *Trematocara*.

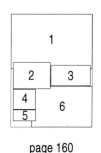

page 160

1. The coast at Chimba, Zambia.
2. Crabs are quite common in the intermediate habitat.
3. *Aethiomastacembelus ellipsifer* is probably the most attractive eel in the lake.
4, 5 and 6. *Auchenoglanis occidentalis* is a large catfish seen mostly in the shallow intermediate habitat and one of the few fishes not endemic to the lake.

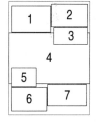

page 161

1. *Neolamprologus modestus* at Mamalesa Island, Tanzania.
2. *Neolamprologus petricola* at Cape Tembwe, Congo.
3. *N. petricola* at Milima Island, Congo.
4. *N. modestus* at Chisanse (Nkamba Bay), Zambia.
5. *Neolamprologus mondabu* at Magara, Burundi.
6. *N. mondabu* at Kapemba, Tanzania.
7. *N. mondabu* at Kekese, Tanzania.

The intermediate habitat

The intermediate habitat consists of a sandy bottom with numerous rocks which provide shelter for the species-rich community. The rocky part can cover up to three quarters of the sandy bottom. The most important characteristic is the gradual inclination of the bottom. The biocover on the rocks is usually overlaid with a thin layer of fine sand. There is no real depth restriction for this habitat, but it is most heavily populated between 5 and 40 metres. The intermediate habitat harbours the most species-rich communities of the lake.

Neolamprologus modestus types

Neolamprologus modestus and *N. mondabu* have confused ichthyologists in the past (Poll, 1956), but these two species are easily identified underwater. *N. modestus* is a very dark brown-coloured cichlid, while *N. mondabu* usually has very light coloration. It may be difficult morphologically to differentiate this pair, but it is even more difficult to distinguish *N. modestus* and *N. petricola*, which look almost identical underwater, in coloration as well as in behaviour. In fact it is not absolutely certain that *N. petricola* does in fact represent a different species to *N. modestus*. Both *N. petricola* and *N. modestus* have an elongate shape, but *N. modestus* has a shallower body. Both species have dark brown coloration, but under suboptimal conditions both show a much lighter pattern. *N. modestus*, in such circumstances, has a much yellower tint than the beige-greyish-coloured *N. petricola*. The submissive colour pattern of the latter consists of broad grey vertical bars on an almost white background. *N. modestus* is characterized by bright yellow pectoral fins; the pectorals of *N. petricola* are colourless or greyish, but never yellow. Large specimens of *N. petricola* exhibit a cranial gibbosity which has never been seen in *N. modestus*. All three species can attain a maximum total length of about 12 cm.

As might be expected, these two species do not have an overlapping range. *N. modestus* occurs between Moliro Bay, Congo, and Kala, Tanzania, and is common in Zambian waters. *N. petricola* is found north of the Lunangwa River in the Congo, at least as far as the Kavala Islands, but may also be found further north. *N. mondabu*, on the other hand, has a lake-wide distribution but is much rarer in areas where either of the other two species is found; it is more common in the northern parts of the lake than *N. modestus* or *N. petricola* are in their ranges. A single geographical variant of *N. mondabu* has been found, at Cape Korongwe in Tanzania. Both male and female have a black stripe on the lower edge of the tail. The tail also has pointed tips, a feature usually seen in very large adults, and not seen in the other two species.

All three species are invertebrate-feeders that feed mostly from the sandy substrate in areas characterized by small rocks and pebbles. A part of the diet consists of small snails — there are a few molariform teeth on the pharyngeal jaws in all three species — but the main food is insect larvae, which they find in the sand by diving into it, either sifting it through the gills or wriggling the body so much that the prey is exposed (Hori, 1983). Molluscs and insect larvae are easily digestible, and the intestinal tracts of these fishes have a length of about 70% of the standard length.

The snails eaten by these three cichlids are juvenile specimens — adult snails of most species in the lake attain a diameter of about 1.5 cm. The snails are ingested whole and crushed between the pharyngeal jaws, the edges of which are packed with long slender teeth that "rake" the liberated soft contents of the mollusc into the gullet. The main problem of snail-crushers lies in the disposal of the ingested shell fragments. Sometimes the fish manages to remove the fragments from the meat and spit them out, but most of the small snails are eaten completely.

Another food source is supplied by *Lamprichthys tanganicanus*, the beautiful killifish of the lake. It deposits its eggs in small cracks in the rocks. Males are territorial and defend a crack in a rock against conspecific males. A ripe female is led to the crack where she shoots a few eggs inside the crack. The male positions himself alongside her and releases his seminal fluid simultaneously. *N. mondabu* is often found following such mating pairs, trying to steal the eggs from inside the crack. The killie does not seem to be disturbed by this robbery, and appears to regard the cichlid(s) as "part of the furniture", as males rarely chase cichlids away from their territories.

Males of all three species have small harems, with two to five females residing within their territories. It is not uncommon to find that the females of *N. petricola* and *N. modestus* are larger than the males. It is the female who digs the nest, under a rock, and tries to attract the male to this nest. The eggs are deposited inside the nest, on the ceiling or against the walls of the cave, and closely guarded by the female. The clutch size can range between 200 and 500 eggs. The free-swimming fry remain close to the substrate in *N. mondabu* and *N. modestus*, but in the aquarium *N. petricola* fry form a school and hover above the nest.

«Lamprologus» caudopunctatus and others

«*Lamprologus*» *caudopunctatus* is one of the most abundant cichlids in the intermediate habitat, and at many places, in particular in Zambian waters, large plankton-feeding schools can be seen. This species occurs in shallow water as well as at depths of more than 25 metres. In the wild it does not grow much larger than 6 cm. In morphology it is almost indistinguishable from «*L.*» *leloupi*, with which it shares the habitat along the Congolese shore between Lunangwa and Kapampa. Where both species are found together «*L.*» *caudopunctatus* has an orange dorsal fin; in other parts of its range «*L.*» *caudopunctatus* has either a yellow-coloured dorsal or the fin is colourless. Büscher (1992a) compared both species and found that «*L.*» *leloupi* has a somewhat deeper body and a pattern of large, chessboard-like, blotches on the anterior half of the body. The basic melanin pattern in «*L.*» *caudopunctatus* consists of vertical bars on the anterior half of the body. These patterns are seen only in distressed individuals or at night, when the fish is at rest.

«*L.*» *caudopunctatus* occurs between Kapampa in the Congo and Kala in Tanzania, and is thus found in Zambian waters. The distribution of «*L.*» *leloupi* is split into two segments on opposite shores of the lake; it is found between Cape Tembwe and the Lunangwa River in the Congo, and between Kasoje and Sibwesa along the central Tanzanian shore. Between Sibwesa and Kala there is a third species with a close resemblance: «*L.*» sp. "caudopunctatus kipili", which occurs in the intermediate habitats between Wampembe (Fulwe Rocks) and Kipili. This fish has a shallower body and a longer snout than either «*L.*» *caudopunctatus* or «*L.*» *leloupi*. There are no geographical variants known, and it appears to be much less abundant than the other two cichlids. The diet of these three species

page 164

1, 2, and 4. The killifish *Lamprichthys tanganicanus* breeds in the rocky habitat. The female shoots the eggs into a crack in a rock (1 and 4) while the male fertilizes them as soon as they leave the female. *Telmatochromis temporalis* and *Neolamprologus mondabu* are notorious egg robbers that investigate the spawning site of the killifish as soon as the pair has vacated it (2).
3. «*Lamprologus*» sp. "caudopunctatus kipili" at Kisambala, Tanzania.
5. A foraging school of «*L.*» *caudopunctatus* at Cape Chaitika, Zambia.

page 165

Geographical variants of «*L.*» *leloupi*-like cichlids and their ranges.
Note that «*L.*» *leloupi* has a disjunct range but is found on both sides of the lake.
Map: black = «*L.*» *leloupi*; red = «*L.*» *caudopunctatus*; blue = «*L.*» sp. "caudopunctatus kipili".

① *«L.» leloupi*, Cape Tembwe

② *«L.» leloupi*, Kanoni

③ *«L.» leloupi*, Kapampa

⑨ *L.» leloupi*, Lumbye

⑧ *«L.» leloupi*, Sibwesa

⑦ *«L.» caudopunctatus*, Mamalesa Is.

⑥ *«L.» caudopunctatus*, Kasanga

③ *«L.» caudopunctatus*, Kapampa

④ *«L.» caudopunctatus*, Moliro

⑤ *«L.» caudopunctatus*, Chituta

consists of all kinds of invertebrates which are picked up from the substrate or from mid-water.

Breeding in all three species takes place in holes dug in the sand under rocks or between small rocks. In many places small rocks are buried under a few centimetres of sand, and such spots are attractive to many cave-brooders. The sand is excavated from between adjacent stones and a network of cracks and clefts becomes available as a nesting site. If such sites are not available the fish may instead pile up a heap of sand against a larger rock and dig a hole in this heap alongside the rock. In captivity they sometimes accept empty snail shells as a breeding caves, but in the wild this is rare.

Clutch size for «*L.*» *caudopunctatus* and «*L.*» *leloupi* ranges between 100 and 250 eggs. When the fry swim free they form a school and hover above the nest, feeding on plankton. Juveniles do not stay with their parents for long, and have to leave the nest before a new brood is produced. Juveniles of previous spawns have never been seen around the nest when a pair is guarding a fresh batch of fry. The parents swim above the fry when they are guarding and chase all intruders from the nest.

«*Lamprologus*» *finalimus* was described from a single specimen which was reportedly collected near Uvira in the Congo. Melanie Stiassny (pers. comm.) found that this specimen was very similar to «*L.*» *caudopunctatus*, but that very little more could be said, based on a single specimen. The area near Uvira has been thoroughly investigated by scores of scientists but they have never found a species resembling «*L.*» *finalimus*. I have been unable to visit the northwestern shore of the lake, and reports on that area, between the Ubwari Peninsula and the Kavala Islands, are sparse. It may be that the range of «*L.*» *finalimus* lies in that area. A single characteristic distinguishes this species from any of the three species mentioned above: the anal fin has a black margin. This feature has not been found in any population of «*L.*» *leloupi* or «*L.*» *caudopunctatus*.

Blue bars

Neolamprologus tretocephalus and *N. sexfasciatus* resemble each other in coloration and in breeding and feeding habits, and are also found sympatrically at some localities. As its name suggests, *N. sexfasciatus* has six dark blue bars on a light blue or yellow-coloured body whereas *N. tretocephalus* has five; the bar between the eyes is not counted. *N. sexfasciatus* is more common in areas that are less sandy than those frequented by *N. tretocephalus*. There are also morphological differences: the pharyngeal teeth of *N. tretocephalus* are stouter than those of *N. sexfasciatus*. Although both species feed primarily on insect larvae and crustaceans, *N. tretocephalus* also feeds on small snails and *N. sexfasciatus* sometimes takes small fish. Poll (1956) found that the intestines of *N. tretocephalus* measure about 30% of the fish's total length while those of *N. sexfasciatus* measure about 90%.

N. tretocephalus may have adapted to eat snails, a food source that most other cichlids are unable to process, but it will eat anything else when available (as will most other species). The most readily accessible food source will normally be exploited first, and only when times are hard (food shortages) do specialist species have to rely on their specialisations in order to struggle through the bad times.

The heavily-built teeth on the pharyngeal jaws help *N. tretocephalus* to crush not only small snails (snails comprise less than 10% of the food taken (Yuma & Kondo, 1997)) but also the shells of mussel shrimps and the cases of caddis fly larvae, food items not often taken by other invertebrate-feeders. In areas where both species are found together, and they are found side by side, they may rely more on their specialised diets than on more generalised fare.

Breeding in both *N. tretocephalus* and *N. sexfasciatus* can be observed throughout the year and breeding pairs guarding their offspring are common. Both species form pairs that stay together for at least as long as there are fry to protect, but both may have a prolonged bond that extends beyond the brood cycle. Both species have been bred in the aquarium, but there seems to be a problem with the southern variant of *N. sexfasciatus* because no one has yet succeeded in spawning them in a tank. A possible reason could be that a pair forms a bond for life and the adult pairs exported for the hobby have become separated.

A nest is dug — in the aquarium it is mainly the female *N. sexfasciatus* that digs the hole — in the sand under a rock or small group of rocks, and the eggs deposited inside the nest, away from view. *N. tretocephalus*, males of which can attain a total length of about 13 cm and females a little less, can produce up to 500 eggs per spawn. *N. sexfasciatus* (maximum total length about 16 cm) is even more prolific and can produce clutches of almost 1000 eggs. The fry are guarded by both male and female; the female stays near the fry all the time while the male guards the perimeter of the territory, which is not very large.

The breeding colour of both male and female differs from the their normal dress of vertical bars. In breeding *N. sexfasciatus* the vertical bars fade except for a section in the middle of each bar, thus forming a mid-lateral row of black spots. This pattern is most striking in the golden-yellow variant from the east coast. In the aquarium both *N. sexfasciatus* and *N. tretocephalus* are very aggressive towards conspecifics and in most cases only a pair can be housed, even in very large tanks.

The range of *N. sexfasciatus* encompasses the entire southern half of the lake, with its northernmost point on the west coast at Cape Tembwe, and on the east coast at Isonga, north of Ikola. There are several geographical variants known. The holotype stems from M'toto in the Congo and the individuals of this population are characterized by a yellow-coloured upper half to the body and head. This variant occurs between Cape Tembwe and Kapampa on the western shore, but a very similar variant is also found at Fulwe Rocks on the opposite side of the lake. Between Kapampa and Livua *N. sexfasciatus* is characterized by a bluish body and a yellowish dorsal fin.

Between Moliro, Congo, and Kala, Tanzania — including the entire Zambian shore — we find the so-called "Blue Sexfasciatus", which has

no yellow pigment. The boundary between the yellow- and the blue-coloured *sexfasciatus* is also the southernmost point in the range of *N. tretocephalus*. Between Cape Tembwe and Livua both species are sympatric, but *N. sexfasciatus* has a yellow cast while *N. tretocephalus*, of which there are no geographical variants known, has a bluish tinge.

On the eastern side of the lake the situation seems to be similar to that on the opposite shore. North of Fulwe Rocks a completely golden-yellow *N. sexfasciatus* is found between Hinde B and Isonga. At Cape Mpimbwe, however, I found two breeding pairs in which the female was yellow but the male blue. All other (single) specimens had yellow coloration, so it may be that the breeding colour of the males in that particular population is blue instead of yellow. Interestingly, Cape Mpimbwe is also the southernmost point along the east coast in the range of *N. tretocephalus*. The latter occurs in a large part of the lake: north of Livua in the Congo along the west coast, and north of Cape Mpimbwe on the eastern side. It is, however, not found in Burundi waters apart from the extreme south near the border with Tanzania.

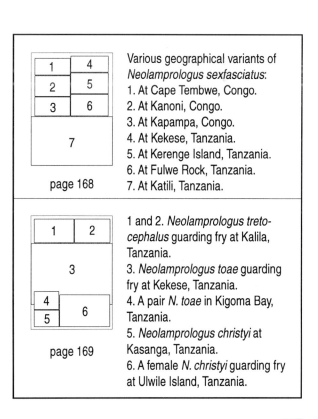

1	4
2	5
3	6
7	
page 168	

Various geographical variants of *Neolamprologus sexfasciatus*:
1. At Cape Tembwe, Congo.
2. At Kanoni, Congo.
3. At Kapampa, Congo.
4. At Kekese, Tanzania.
5. At Kerenge Island, Tanzania.
6. At Fulwe Rock, Tanzania.
7. At Katili, Tanzania.

1	2
3	
4	6
5	
page 169	

1 and 2. *Neolamprologus tretocephalus* guarding fry at Kalila, Tanzania.
3. *Neolamprologus toae* guarding fry at Kekese, Tanzania.
4. A pair *N. toae* in Kigoma Bay, Tanzania.
5. *Neolamprologus christyi* at Kasanga, Tanzania.
6. A female *N. christyi* guarding fry at Ulwile Island, Tanzania.

Neolamprologus toae

Territories of *Neolamprologus toae* are commonly found in close proximity to those of *N. tretocephalus*. Although *N. toae* has a dark brown to black dress when adult, it is nevertheless quite an attractive fish to look at. Maximum total length is approximately 10 cm.

N. toae feeds during the night or in the dim light of dusk and dawn. The relatively short length of its intestine — 60% of the fish's total length (Poll, 1956) — indicates a carnivorous diet, and the diet in practice consists for the most part of shrimps, but other soft invertebrates, such as insect larvae, are eaten as well. Large clouds of shrimps can sometimes be seen in shallow water at dusk and may constitute the food of a number of cichlid species feeding at this time.

N. toae has small indentations above the upper lip and around the eyes, which are enlarged sensory pores that are advantageous in locating prey under subdued light conditions. In fact the sensory system in *N. toae* is enlarged to such an extent that Colombé and Allgayer (1985) placed this species in a genus of its own, *Paleolamprologus*, assuming that this was the primitive state of the sensory system in lamprologines. Stiassny (1997), however, has suggested that, quite to the contrary, the sensory system of *N. toae* is advanced in comparison to that of other lamprologines, but that there are no other characters that would warrant the separation of this species into a monotypic genus.

Given the sharing of the habitat at most locations with *N. tretocephalus*, also a predator on invertebrates, one might conclude that these two species are competitors not only for spawning sites but also for food. Investigations of stomach contents have revealed, however, that the two species have different preferences as regards prey items. It has been found (Hori, 1983) that the diet of *N. toae* comprises more than 80% shrimps, while that of *N. tretocephalus* features mainly insect larvae. There are other shrimp-eaters in the intermediate habitat but none of them seems to be so well-equipped to feed at night as *N. toae*. This species is also much less aggressive than *N. tretocephalus* in defending its feeding territory.

Breeding *N. toae* are normally found in shallow water and both male and female guard the offspring. Nagoshi (1987) found that a pair guard their fry for about 12 weeks before they start a new brood. He found on two occasions that a single male tried to raise two broods with two different females but was unsuccessful: the raising of sufficient numbers of offspring requires parental care by both parents. The eggs are attached to the upper face of a small flat rock (Nagoshi, 1987) or to a vertical surface inside a small crevice or between two adjacent stones. After about four to five days the larvae are transferred to a rocky crevice or left at the bottom of the spawning cave. A few days later the fry are free-swimming and hover above the nest: usually no further away than about 10 cm initially and about 30 cm when they are a few weeks old. Clutch size ranges between 150 and 250 eggs.

The holotype of *N. toae* was collected near Toa on the central Congolese coast, but the species has a much larger range and is found in the entire northern half of the lake, except in Burundi. The southernmost point in its range on the west coast is Cape Tembwe, and on the east coast it occurs as far south as Cape Mpimbwe. There are no geographical variants known.

Neolamprologus christyi

N. christyi is a peculiar cichlid which, at first sight, appears to be closely related to *N. mondabu* but has far more scales on the anterior part of the body. Almost all morphological characters of these two species are in the same range apart from a more forked tail in *N. christyi* and the number of scales. Trewavas and Poll (1952) give 50 to 60 scales in a horizontal row for *N. christyi* while *N. mondabu* has an average of 35. The majority of *N. christyi*'s scales are found on the nape and shoulders where they are much smaller than elsewhere on its body. At Hinde B I found both *N. christyi* and *N. mondabu* in the intermediate habitat so it cannot be argued that *N. christyi* is simply a geographical variant of *N. mondabu*. It must be said, however, that *N. mondabu* from that area have a slightly forked tail as well, but the specimens I saw were smaller and much lighter in colour than *N. christyi*.

N. christyi is rather common in habitats where rocks border larger stretches of sand, normally in water with a depth of less than 10 metres. It is an invertebrate-feeder: stomach contents inventories have included crustaceans, worms, and even small snails. Maximum total length is about 15 cm.

Its range is restricted to the southeastern part of the lake, where it occurs between Isanga in Zambia (Herrmann, 1990) and Kipili in Tanzania. In the northern part of its range it is found in less sediment-rich areas than in the south, where it occurs in sometimes very turbid water near river outlets (e.g. near Samazi).

Juvenile *N. christyi* have a bluish cast over a beige-coloured body, but adults are matt black and difficult to make out in dark caves. Before spawning the pair dig a hole in the sand under a rock and the eggs are hidden in this cave, probably attached to the ceiling. The guarding of the fry is performed mainly by the female. Clutches can contain more than 250 eggs.

Neolamprologus pectoralis and N. nigriventris

In habitat preference, coloration, and morphology, *Neolamprologus pectoralis* to some extent resembles *N. christyi*, but is found on the opposite shore, in the Congo. The holotype of *N. pectoralis* was collected near the village of Tembwe (not to be confused with Cape Tembwe and therefore often called Tembwe Deux); further type material was caught at Kizike, about 20 km south of Tembwe. Büscher (1991) reports that *N. pectoralis* has also been seen south of Lunangwa, about 80 km south of Tembwe, but specimens from this area were not included in the type series. The area in between, a stretch of about 60 km, is inhabited by a closely-related but different species: *N. nigriventris*.

The name *pectoralis* alludes to the large pectoral fins which are a remarkable and conspicuous feature of the species. The length of the pectoral fin is about equal to the length of the head,

i.e. much larger than in other known lamprologines with a rounded caudal fin. Another unusual feature is the length of the lower lateral line, which extends from a little behind the gill cover to the caudal peduncle. The lower lateral line in *N. nigriventris* is also very long and starts about eight scales behind the gill cover. In all other known cichlids the lower part of the system is much shorter and begins on the rear half of the body. Using these two features (length of pectoral fin and lateral line) it is simple to distinguish *N. pectoralis* from other lamprologines. It may be closely related to *N. furcifer*, a cichlid with a similar body shape and long pectoral fins (90% of head length (Poll, 1956)); it is, however, easily distinguished from that species (and also from *N. christyi*) by the rounded caudal fin; that of *N. furcifer* has long filamentous lobes.

N. pectoralis attains a total length of about 14 cm. Females remain a little smaller than males. The maximum total length of *N. nigriventris* is about 10.5 cm and this species is thus noticeably smaller than *N. pectoralis*. The main difference between these two species is, however, the

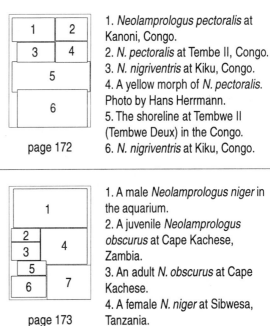

page 172

1. *Neolamprologus pectoralis* at Kanoni, Congo.
2. *N. pectoralis* at Tembe II, Congo.
3. *N. nigriventris* at Kiku, Congo.
4. A yellow morph of *N. pectoralis*. Photo by Hans Herrmann.
5. The shoreline at Tembwe II (Tembwe Deux) in the Congo.
6. *N. nigriventris* at Kiku, Congo.

page 173

1. A male *Neolamprologus niger* in the aquarium.
2. A juvenile *Neolamprologus obscurus* at Cape Kachese, Zambia.
3. An adult *N. obscurus* at Cape Kachese.
4. A female *N. niger* at Sibwesa, Tanzania.
5. *N. obscurus* at Kambwimba, Tanzania.
6. *N. obscurus* at Cape Tembwe, Congo.
7. *N. obscurus* at Kiku, Congo.

length of the pectoral fins and the characteristic coloration of *N. nigriventris* (see photos).

The food found in the stomachs of some specimens of both species comprised crustaceans, insect larvae, copepods, snail fragments, algal strands, and sand grains (Büscher, 1991, 1992d). In the natural habitat *N. pectoralis* maintains close contact with the rocky substrate and moves around the rocks in the same fashion as is seen in *N. furcifer*, i.e. the ventral part of the body remains just a few millimetres above the substrate while the fish follows the contours of the rock.

N. pectoralis is a rather rare cichlid and occurs at depths deeper than 15 metres. Adult pairs have sometimes been observed, probably in front of their caves, but most other individuals seen have been solitary sub-adults. Büscher (1991) found that the nests occupied by breeding pairs were horizontal cracks with a very low entrance but extending deep into the rock (deeper than 60 cm).

In its natural habitat *N. pectoralis* is readily recognised by its large pectorals which are moved slowly but steadily. This species creates a very striking impression, especially when seen from above: it is not so much the size but more the deep colour of the pectorals that makes them so conspicuous. Many lamprologines have colourless pectoral fins but those of *N. pectoralis* have a deep colour. There are two colour morphs of this species known, dark brown and yellow, and the pectorals are coloured accordingly. Büscher notes that both colour morphs are present at all locations, but that the yellow morph is rare. The yellow colour is seen in juveniles as well as in adult individuals.

Some hypotheses about the large pectoral fins of *N. pectoralis* have been given by Büscher, who suggested that on the one hand these fins may give the fish a more accurate control of movement and on the other they may optimize the flow of oxygen-rich water inside the narrow caves they occupy. *N. furcifer* also has large pectoral fins (but faintly coloured) and behaves in a similar way to *N. pectoralis*. So it may indeed be that these fins play a role in the characteristic way these species move about over the substrate. The movements of *N. nigriventris* appear less typical to me and its pectoral fins are within the size range of most other lamprologines.

As already mentioned, another feature which has no equal among other Tanganyika cichlids is the highly developed lateral line system of both species. This may increase the sensitivity of predator detection (in which case it should have developed in other cichlids as well), but it is more likely an optimized method of prey detection.

When observing *N. pectoralis* in its natural environment it has been noted that, except for the pectoral fins, it does not move much. The movement of invertebrate prey (crustaceans) may be stimulated by the flow of (oxygen-rich) water. The continuous paddling with the fins may thus arouse invertebrates in the immediate vicinity of the fish. Their movements can be detected by the lateral line system and, thanks to the extension forwards, also close to the cichlid's head. Other cichlids (e.g. *Trematocara*) have highly-developed sensory pores on the head with which they detect prey, and it may be that *N. pectoralis* instead developed the other part of the sensory system for prey detection. Such paddling behaviour has not been seen in *N. nigriventris* although its lateral line system is also enlarged.

Dark and obscure cichlids

Neolamprologus niger and *N. obscurus* are both shy, dark-coloured cichlids that live in the crevices of the sediment-rich intermediate biotope. They are usually found at a depth of between 6 and 30 metres and are among the most secretive cichlids of the lake.

The holotype of *N. niger* was caught at Luhanga in the extreme north of the lake, but Poll (1956) mentions in his description that it was also found in Kungwe Bay, Tanzania. The range of *N. niger* encompasses the area between Sibwesa in Tanzania and the Kavala Islands in the Congo, but it is absent from Burundi waters (Brichard, 1989). In its preferred habitat, i.e. small rocks and lots of sediment and sand, *N. niger* is a rather common species.

N. obscurus seems to be the counterpart of *N. niger* in the southern section of the lake, but has not (yet) been found along much of the Zambian and Tanzanian coast. Its range lies between Cape Tembwe and Cape Kachese, and I also found a small population just north of the Kalambo River in Tanzania. There are no reports of *N. obscurus* from the area between Cape Kachese and the Kalambo River.

The diet of these two little cichlids — the maximum total length of *N. niger* is about 9 cm and that of *N. obscurus* approximately 8 cm — consists mainly of insect larvae and crustaceans, but tiny snails are sometimes eaten as well.

Juveniles of *N. obscurus* are light brown with a pattern of a series of double bars; adults are entirely dark brown and the pattern is only partly visible. Juvenile *N. niger*, however, are orange, and this colour, usually a little more yellow, is also seen in sexually active females. Old and large females, however, become dark-brown like the adult males. The entire life style of these two cichlids is secretive and thus also their spawning site. In the aquarium *N. niger* breeds regularly with very small broods, sometimes no more than three or four fry. The fry of previous spawns are tolerated inside the breeding cave. *N. obscurus* has larger spawns but usually no more than about 20 fry per brood. The pair do not directly guard their offspring but rather the spawning cave; if the fry venture too far from the nest then they are literally on their own.

An uncommon characteristic, as far as lamprologines are concerned, is the enlarged sensory pores on the gill cover and jaws of *N. niger*. Such pores could be helpful in locating invertebrates in the dark corners of its environment. The idea that such enlarged pores have developed for an enhanced and early detection of predators is somewhat contradicted by their location on the fish. If a more sensitive warning system were necessary the fish would probably have developed an extra lateral line, as it is likely that the total body of the fish acts as a better (but less precise) receiver than just the head. This hypothesis is supported by the nature of the sensory system in many cichlids that spend most of their time on the open sand floors of the lake. Members of the genus *Xenotilapia*, for example, are characterized by three lateral lines on their sides (not on their head). The cephalic pores, on the other hand, permit their owner precisely to locate prey even if it is not visible. In the case of *N. niger* the prey might perhaps be hidden in the darkness of its refuge; for other species, e.g. *Aulonocranus*, prey is hidden in the sand.

Yellow Julies

The genus *Julidochromis* in this biotope is represented by two species: *J. regani* and *J. ornatus*. *J. regani* is the only "Julie", as members of this genus are commonly known among aquarists, that can be encountered on a pure sandy bottom, but always near rocks. Brichard (1978) reports that they tend to flee away over sand rather then seeking refuge among the rubble. However, in most cases *J. regani* stays close to the rocky substrate and Brichard may have been referring to the population in the northernmost part of the lake, near Bujumbura in Burundi, where the intermediate habitat contains only very few rocks.

J. ornatus and *J. regani* are a much yellower colour than *J. marlieri* and *J. dickfeldi*, and this may be some kind of adaptation to the sandy habitat in which they two are normally found. Usually all Julies occur in crevices and caves of the rocky substrate, and *J. regani* and *J. ornatus*

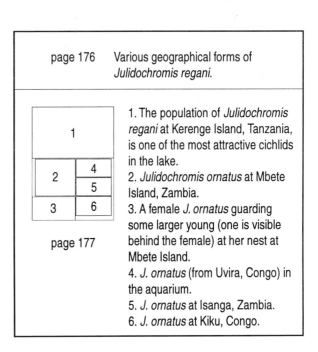

page 176   Various geographical forms of *Julidochromis regani*.

1. The population of *Julidochromis regani* at Kerenge Island, Tanzania, is one of the most attractive cichlids in the lake.
2. *Julidochromis ornatus* at Mbete Island, Zambia.
3. A female *J. ornatus* guarding some larger young (one is visible behind the female) at her nest at Mbete Island.

page 177

4. *J. ornatus* (from Uvira, Congo) in the aquarium.
5. *J. ornatus* at Isanga, Zambia.
6. *J. ornatus* at Kiku, Congo.

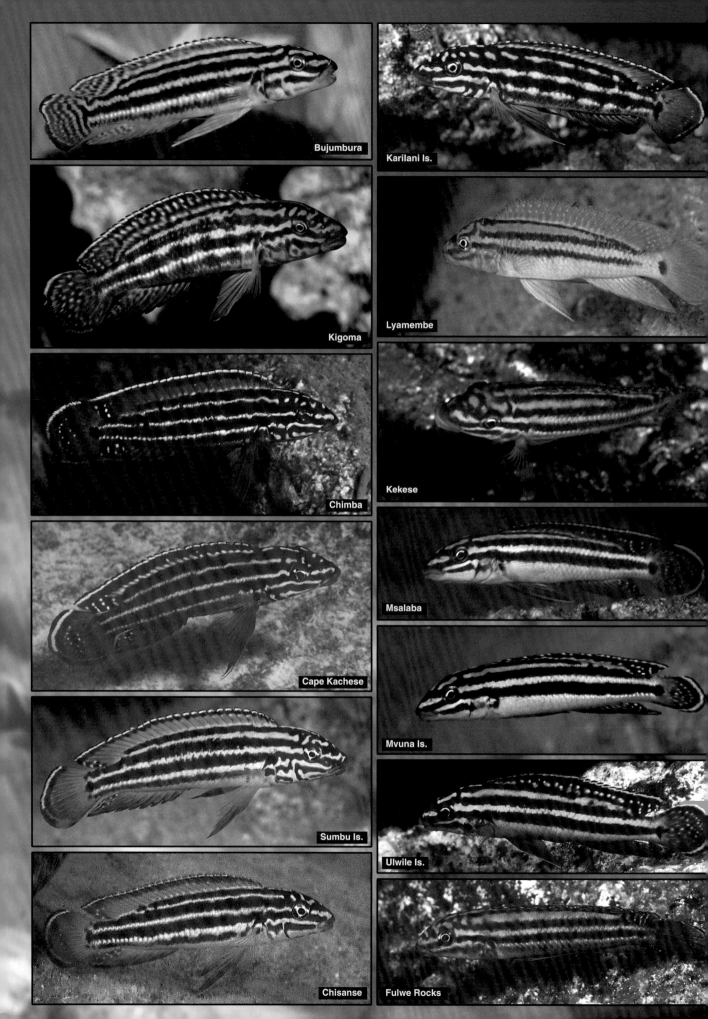

Bujumbura

Karilani Is.

Kigoma

Lyamembe

Chimba

Kekese

Cape Kachese

Msalaba

Sumbu Is.

Mvuna Is.

Ulwile Is.

Chisanse

Fulwe Rocks

are at home in habitats where sand and sediment cover part of that substrate.

Breeding takes place inside narrow clefts in rocks or, more commonly among the yellow Julies, under (flat) rocks that are a few centimetres above the sand. As in *J. marlieri*, females of *J. regani* are noticeably larger than males. In *J. ornatus* the males are larger. Males can breed with several females but in *J. regani* it is the female that decides where to spawn. In *J. ornatus* the offspring are guarded by both parents although their spawns are much smaller (about 25 eggs per clutch) than those of *J. regani*, which can have broods of more than 150 fry. Fry of previous spawns are tolerated in the territory but are chased away from the eggs of a new brood, and are expelled from the territory when they have reached a length of about 3 cm.

J. regani has the widest range of all Julies (see map page 105), occurring mostly along the eastern and southern shores of the lake. It occurs in the extreme northeastern corner of the lake, near Bujumbura; south of Resha, Burundi, across the border into Tanzania to the Malagarasi River mouth; between Bulu Point and Fulwe Rocks (Wampembe) in Tanzania; and between Nkamba Bay and Katete in Zambia. Brichard (1978) records a population around the Ubwari Peninsula. Many geographical variants are known and quite a number of these are kept by aquarists.

The holotype of *J. regani* was collected at Nyanza-Lac in Burundi. At various places around the lake there are *J. regani* populations consisting of dark-coloured individuals with a pattern consisting of the usual horizontal stripes, but, in addition, these stripes are thickened at certain intervals coinciding with the position of vertical bars in *J. marlieri*. Such populations occur between Resha and Rumonge in Burundi, between Kigoma and the Malagarasi River, at Karilani Island, and at Ubwari. Brichard (1978) described this form as *J. regani affinis*, a subspecies of *J. regani*. There seems to be a lack of stable characters distinguishing *J. regani* from *J. marlieri*, a situation not unlike that found between *J. ornatus* and *J. transcriptus*. And as in the case of the latter pair, *J. regani affinis* is always found between popula-

tions of *J. marlieri* on the one side and *J. regani regani* on the other. One could argue that the presence of vertical bars, making the fish less obvious, is an adaptation to living in a rocky environment, and that *J. regani regani*, found in open intermediate habitats, has a better survival rate when it is not dark-coloured and thus lacks vertical bars. Although this seems to be the case in most situations there are dark-coloured *J. regani* (and *J. transcriptus*) found in open and light environments. Brichard (1978) also mentions a large population of *J. marlieri* in a shallow intermediate habitat at Ruziba (Burundi).

I have been able to identify with some confidence all populations of *J. regani* and *J. marlieri* that I have seen so far, and there seems to be, on the basis of current evidence, no need to retain subspecific status for the darker *J. regani*. All populations of Julies with vertical bars and with a black stripe below the eye are *J. marlieri* and those without continuous vertical bars are *J. regani*.

The populations found along the central Tanzanian shores, between Kasoje and Fulwe Rocks, are characterized by two, instead of three, horizontal lines on the body. Such a character is sufficiently distinct to assign these populations to a different species, if one so desires, but I regard them as merely geographical variants of *J. regani*.

Julidochromis ornatus was described by Boulenger (1898) from a population which inhabits the rocky shores near Mpulungu. This population is characterised by individuals with a white-yellow ground colour, two black horizontal stripes, and a black band in the lower part of the dorsal fin. They have been exported under the trade name of "White Ornatus". Matthes (1962) re-described *J. ornatus* on the basis of specimens found in the northern part of the lake, near Uvira. Three years earlier the same author described *J. transcriptus* from a nearby area (see discussion page 102).

Around the Kapampa-Kileba area in the Congo a similar situation is found to that described by Brichard (1978, 1989) for *J. transcriptus* and *J. ornatus* in the north of the lake. The same two species live in the intermediate habitat and are geographically separated: at

Kapampa *J. transcriptus* occurs over a considerable depth range and also exhibits a wide range in coloration, and north and south of Kapampa we find *J. ornatus* (characterized by horizontal stripes and lack of vertical bars); both *J. transcriptus* and *J. ornatus* lack the black stripe under the eye. Brichard (1978, 1989) reports that *J. transcriptus* can be found in different populations separated by just a few miles of sandy beach and alternating with populations of *J. ornatus*. *J. transcriptus* and *J. ornatus* have not been found sympatrically, which is also true for *J. marlieri* and *J. regani*.

The known distribution of *J. ornatus* is divided into three geographically separated sections of the lake: near Uvira in the extreme north of the lake; between Mpala and Lunangwa Bay; and between Kasakalawe, a few kilometres west of Mpulungu, and Kantalamba, just across the Kalambo River mouth into Tanzanian waters. See distribution patterns of all Julies on page 105.

Lepidiolamprologus lemairii

After a very comprehensive and meticulous morphological study, Stiassny (1997) found that *Lamprologus lemairii* should be assigned to *Lepidiolamprologus*, a suggestion which is followed here.

L. lemairii is a common predator of the intermediate habitat and occurs in shallow as well as in deep water. The maximum total length this piscivore can attain is about 25 cm, at least for males; females remain approximately a third smaller.

L. lemairii feeds on small fishes but details of its feeding behaviour are not known. Hori (1983) assumes that it feeds at night, as during the day feeding activity of *L. lemairii* is very rare. He found that the stomach contained food in the morning but was empty in specimens caught late in the afternoon. This predator behaves rather lethargically during the day, but this is probably part of its hunting technique. Its coloration, consisting of an irregular pattern of light and dark blotches, resembles that of several other cichlids in other parts of the world which also practise an ambush type of hunting behaviour, e.g. *Nimbochromis polystigma* from Lake Malawi, *Haplochromis cavifrons* from Lake Victoria, *Serranochromis longimanus* from the Okavango Delta, one of the morphs of *Herichthys minckleyi* from Mexico, and *Crenicichla* sp. *aff. jegui* from Brazil. The cryptic coloration of these predators allows them to get near the fish on which they prey. By remaining motionless they virtually blend in with the environment and other fishes are practically unaware of their presence. I have observed on several occasions that *L. lemairii* selects its "resting" area close to groups of juvenile cichlids. As it lies motionless on the substrate, the inquisitive young fish may come near to the predator, sometimes close enough for it to strike.

L. lemairii is a hole-brooder, in which females choose holes which are just large enough for them but too small for the male to fit in. Males are seen near the nest just before and after spawning, but almost all guarding is done by the female. The eggs are attached to the rocky substrate inside the hole and the entrance to the nest is blocked off by the female's body. The fry are protected as long as they stay inside the nest but are not retrieved by the female once they wander away.

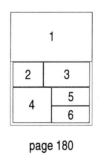

1. *Lepidiolamprologus lemairii* at Mamalesa Island, Tanzania.
2. A juvenile *L. lemairii* sheltering in an empty shell at Mbete island, Zambia.
3. *L. lemairii* at Magara, Burundi.
4. A brood-guarding pair of *Lepidiolamprologus attenuatus* at Kambwimba, Tanzania.
L. attenuatus at Isanga, Zambia (5) and at Cape Tembwe, Congo (6).

page 180

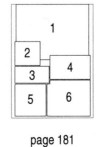

1. A fry-guarding female *L. attenuatus* at Kapemba, Zambia.
2. A typical spawning site of *L. attenuatus* (M'toto, Congo).
3. *L. attenuatus* at Kipili.
4. A female *Lepidiolamprologus* sp. "meeli kipili" at Kipili, Tanzania.
5 and 6. An empty shell in the rim of the nest serves as a shelter for fry of *L.* sp. "meeli kipili", at Kipili.

page 181

L. lemairii is found throughout the lake but geographical variation is not known to occur in this species.

Slender predators

Lepidiolamprologus attenuatus is a very common inhabitant of the shallow intermediate habitat. It feeds on small fishes and invertebrates, and is usually found in small groups or pairs. *L. attenuatus* is characterized by a black spot on the middle of its body, which is particularly prominent in breeding individuals. Its maximum total length is about 14 cm.

Before spawning the pair excavate a small nest between some stones on the bottom. Eggs are stuck onto the vertical sides of these stones and are visible from the outside. After the eggs have hatched and the fry are free-swimming (after about five days), the parents lead the fry through the habitat. During the first few days the fry are mobile they hover above the nest and are fanatically guarded by both parents. Conspecifics are commonly found among predators feeding on fry a few days old.

Lepidiolamprologus sp. "meeli kipili" was at first erroneously introduced as *Neolamprologus meeli* (Konings, 1995), which explains the provisional name of this slender predator, whose maximum total length is estimated at about 11 cm. The "Meeli Kipili" occurs in the sandy part of the intermediate habitat and breeding takes place around an empty snail shell. Both male and female are too large to fit comfortably inside an empty *Neothauma* shell, so this kind of shelter can be used only when the fishes are immature. However, the *Neothauma* shell plays an important role during the breeding procedure of this species. The eggs are deposited on the outside of the shell and protected by both parents. After they have hatched the wrigglers are transferred into the shell. The empty shell is concealed by spraying sand over it. Apart from hiding the shell in the sand, *L.* sp. "meeli kipili" digs a small crater in front of its entrance. The end product is a concealed shell buried in the rim of a crater nest (see photos page 181).

The breeding coloration resembles that of *L. attenuatus* — in fact it may even be a geographical variant of *L. attenuatus* — and it is rather difficult to distinguish these two species. *L. attenuatus* grows to a larger size and has a different breeding behaviour. Sub-adult *L. attenuatus* can be distinguished from adult Meeli Kipili by having a longer snout and shorter pelvic fins. Both species (forms?) are found over sandy substrates in relatively shallow water (depth range extends from 5 to 20 metres) and can be found side by side. As mentioned earlier the Meeli Kipili uses an empty shell to protect the fry, whereas *L. attenuatus* deposits the eggs on the vertical surfaces between small rocks. The latter species is therefore found near small rocks; the Meeli Kipili, however, was found (near Kipili) on the open sand floor, usually near large sand-nests (abandoned) of *Oreochromis tanganicae*.

Brood-guarding Meeli Kipili pairs stay close to the substrate and their offspring forage from the plankton which the water current brings right in front of the shell entrance. Brood-guarding *L. attenuatus* pairs hover above their plankton-feeding brood and are usually seen 60 to 100 cm above the substrate. Their broods usually contain more than 150 fry whereas the number in Meeli Kipili broods appears to be barely more than 50.

The Meeli Kipili has been seen only near Kipili, and could represent a local form of *L. attenuatus*, although specimens with a typical *L. attenuatus* coloration were seen nearby (see photo page 181). Its peculiar breeding behaviour may be an adaptation towards the more open habitat where small stones, the preferred spawning substrate of *L. attenuatus*, are rare, so that the "Meeli Kipili" may have adopted empty snail shells as a breeding receptacle instead.

Thick lips

Lobochilotes labiatus spends most of its life wandering about on its own and probing pockets of sediment and debris with its large fleshy lips. It can attain a maximum total length of about 35 cm and features among the most coveted food fishes of the local fishermen. The mature male does not resemble the silvery-striped juveniles which are rather common in this habitat. Adult male *L. labiatus* have a coloration of irregular orange and green patches and stripes; only traces of the bar pattern are seen. Adults are extremely shy and often found in the extreme shallow water where they hide in rocky caves. *L. labiatus* is a maternal mouthbrooder and the female holds the fry in her mouth, when necessary, several weeks after they have been released for the first time.

The most remarkable characteristic of this species is the protruding, fleshy lips. As in other thick-lipped cichlids, such lips are designed to seal off the buccal cavity during feeding. The lips are very soft and flexible and when they are pressed against the substrate, as the fish feeds, they fit themselves exactly to the contours of the substrate, creating a perfect seal. Where the substrate has a very coarse surface *L. labiatus* needs thicker and more flexible lips to be able to create a seal, and in areas with rough-surfaced rocks the lips of the local *L. labiatus* are actually larger than those of individuals in areas where the rocks have a smoother surface. The growth of the lips may be stimulated, like the callosity of one's hands, by frequent contact with coarse rocks.

It appears that *L. labiatus* does not directly search for prey, but rather for sites where prey may be hidden (Yamaoka, 1997), e.g. pockets in rocks, or interstices between rocks, which are filled up with sediment or debris. Its diet consists mainly of crustaceans and insect larvae.

«*Gnathochromis*» *pfefferi*

«*Gnathochromis*» *pfefferi*, a maternal mouthbrooder, can grow to a maximum total length of about 14 cm and is a common predator of the intermediate habitat. In fact one can encounter this species in any type of habitat. Juveniles are sometimes found in schools among the weeds of the shallows. «*G.*» *pfefferi* is specialized in feeding on shrimps and more than 80% of its diet consists of this type of crustacean (Yuma & Kondo, 1997).

In dense populations males stake out territories, but when only a few sexually active males are in the area, spawning can take place at any site and males appear non-territorial. This type of breeding territoriality is common practice among the haplochromines of Lake Malawi. Sneak males (males that sneak in and "steal" a spawning with a female while she is mating with the "principal" male) are rather common, and I have witnessed one infringing the spawning rights of a non-territorial male. Only females brood the eggs and larvae, and sometimes several mouthbrooding females can be seen together.

The type species of the genus is *G. permaxillaris*, and if one observes both species currently assigned to this genus in their natural habitat, it is not difficult to see that they are very different and unlikely truly to be members of the same genus. *G. permaxillaris* is a *Limnochromis*-like cichlid which has a biparental mouthbrooding breeding technique, while «*G.*» *pfefferi* is more of a *Haplochromis*-like species. According to Poll (1981) «*G.*» *pfefferi* cannot be assigned to a haplochromine genus because its pharyngeal apophysis does not conform to that found in members of that group. For this reason I have placed the name between chevrons.

Petrochromis

P. fasciolatus stands apart from the other species of the genus by virtue of its protruding

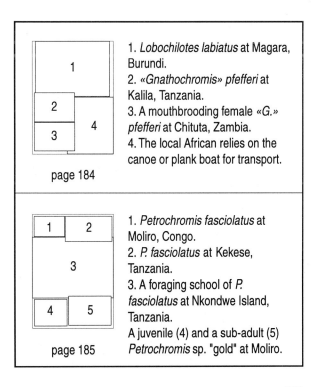

1. *Lobochilotes labiatus* at Magara, Burundi.
2. «*Gnathochromis*» *pfefferi* at Kalila, Tanzania.
3. A mouthbrooding female «*G.*» *pfefferi* at Chituta, Zambia.
4. The local African relies on the canoe or plank boat for transport.

page 184

1. *Petrochromis fasciolatus* at Moliro, Congo.
2. *P. fasciolatus* at Kekese, Tanzania.
3. A foraging school of *P. fasciolatus* at Nkondwe Island, Tanzania.
A juvenile (4) and a sub-adult (5) *Petrochromis* sp. "gold" at Moliro.

page 185

lower jaw. Whereas other *Petrochromis* have a terminal or slightly downwards-pointing mouth, *P. fasciolatus* has an upward-pointing mouth opening. The name *fasciolatus* alludes to the prominent vertical barring found in juveniles and females. Males, which can attain a maximum total length of 15 cm, have a green-grey or blue-grey cast on top of the barred pattern. One population along the central Tanzanian shore sports males with a very bright orange-red eye. This geographical variant has been infrequently exported under the trade name of "Petrochromis Red Eye".

Like all other *Petrochromis*, *P. fasciolatus* feeds on algae (diatoms) which are combed from the biocover on rocks. Since *P. fasciolatus* is found mainly in intermediate habitats, such biocover is often covered with detritus and sand. Stomach contents analyses have revealed that large quantities of sand are ingested together with diatoms (Yamaoka, 1997). *P. fasciolatus* feeds from the vertical faces of rocks in an effort to try and avoid eating too much sediment. Yamaoka (1997) found that *P. fasciolatus* has the lowest number of teeth when compared to other *Petrochromis* species, but in order to compensate feeds with the highest number of bites per second.

In some areas *P. fasciolatus* is very abundant and large foraging schools are regularly seen moving about in adjoining rocky habitats, feeding from the biocover within the territories of other, larger and more aggressive, *Petrochromis* species. A single individual would be unable to feed in such places, but force of numbers makes it possible. Feeding schools of this type are found in other algae-feeding species as well, e.g. *P. famula* and *Tropheus moorii*.

Territorial males are rare, and generally defend spawning sites in the somewhat deeper regions of the intermediate habitat, at depths ranging between 10 and 15 metres. Mouthbrooding females are normally seen on their own, hiding among the rocks, but sometimes they group together and are found in more open environments.

Petrochromis sp. "gold" is regularly seen in the intermediate zone, and also in the pure rocky habitat. I believe that the golden yellow colour

of juveniles and sub-adults is an adaptation towards the sand-rich environments in which they are mostly seen. It is possible that they acquire the yellow pigment by feeding on special types of algae which grow mainly in intermediate habitats, because in the aquarium they tend to lose this colour, even when they are still young. Upon maturation *P.* sp. "gold" loses the yellow colour (Herrmann, 1994b), but I have been unable to find such specimens in the wild.

The king of the featherfins

The featherfin cichlids of the genus *Cyathopharynx* are among the most striking cichlids of Lake Tanganyika. The iridescent colours of territorial males, the building of large sand-castle nests, and the fact that these species are regularly observed while snorkelling, makes them very popular with visitors to the lake. *Cyathopharynx* have been exported from the lake for many years, and since the early eighties geographical races from Tanzania and Zambia, and later from the Congo, have reached hobbyists' aquaria. Even though they are delicate, difficult-to-ship, fishes, and despite the fact that the males lose their vibrant coloration shortly after they have been captured, they have become very popular among aquarists.

Cyathopharynx feed on the unicellular algae and diatoms which are present in the biocover on rocks as well as in the sediment layer on the sand. Non-breeding *Cyathopharynx* usually school together in foraging groups and suck, rather than browsing or grazing, the algae-containing sediment from the substrate. Their teeth seem to be inadequate to pick out specific food items (Yamaoka, 1997), and they therefore have to consume copious amounts. The low nutrition value of the ingested material causes an almost continuous and noticeable flow of excrement. The intestines are about three times as long as the fish itself and contain large amounts of fine sand together with the digestible material (Poll, 1956).

Males can attain a total length of about 22 cm but adult females are normally 12-15 cm. Males are territorial and build sand-castle nests on flat rocks or on the sand. Usually several territorial males can be found together in breeding colonies (leks). Ripe females enter the breeding arenas of the males and may spawn with several of them. A male leads a willing female to the centre of his nest and shows her where to lay her eggs by dragging his ventral fins over the spawning site. He may discharge seminal fluid at this time. While the female deposits some eggs the male waits nearby, sometimes hovering a little above the female. As soon as the eggs appear the male fertilizes them by swimming over them while they are still on the sand. The female turns around and picks up the fertilized eggs, and repeats the spawning sequence. Mouthbrooding females are normally found together in groups and probably release their offspring simultaneously.

C. furcifer was first described by Boulenger in 1898 as *Paratilapia furcifer*. A year later he described *Tilapia grandoculis*, a species which was placed in synonymy with *C. furcifer* years later by Poll (1946). Vaillant (1899) described *Ectodus foae*, which was placed in synonymy with *grandoculis* by Boulenger but which turns out to have an earlier publication date than *C. grandoculis*. Martin Geerts (pers. comm.) has further found that Vaillant later corrected the name to *foai* because the fish was named in honour of a Mr. Foa.

Although it has subsequently become apparent that there are several different geographical races of *Cyathopharynx* there has never previously been any suggestion that more than one species could be involved. Not until I found two different forms of *Cyathopharynx* breeding side by side in Moliro Bay, Congo, did I realise that there must be at least two species in this genus. As the females of the two species are almost indistinguishable, it has been rather difficult to discover which of the two species in Moliro Bay is conspecific with the holotype of *C. furcifer*.

C. furcifer was described using sub-adult specimens collected near Mpulungu. Boulenger gives the following colour description of the holotype: "Bluish above, white beneath; a few ill-defined yellow streaks along the body; some yellow marblings on the postocular part of the head; fins white, with some yellow streaks on the dorsal and anal, and between the ventral

and caudal rays." From this description it is evident that Boulenger is describing the gold-headed *Cyathopharynx* and not the darker species. Furthermore, in alcohol preserved *C. furcifer* have clear dorsal and anal fins; while males of the other form, which I have previously named *C.* sp. "dark furcifer", have dark-coloured fins, again in the preserved state.

In his description of *Tilapia grandoculis* Boulenger gives the following colour description: "Brown above, with ill-defined darker spots, whitish beneath; pectorals yellowish, other fins blackish towards the end." After several observations of mixed groups of *Cyathopharynx* in the wild I found that *C. furcifer* females have an indistinct pattern of three to four blotches on the flank and females of the dark species have somewhat dark-edged fins in addition to a pattern of an unclear blotches on the flank.

After examining a male of each species, collected at Isanga, Zambia, I was unable to find a clear distinction between the two; the main differences appear to be the colour of the fins and that the dark species has a slightly deeper body. These characters coincide with those given for *C. foai* and *T. grandoculis* and the "Dark Furcifer" is therefore here regarded as *Cyathopharynx foai*.

In contrast to females, males of both species are easily told apart. *C. furcifer* from Zambia have a golden patch on the head and light-coloured dorsal and anal fins. The territorial coloration of *C. foai* is greenish or bluish marbled all over the body; furthermore the dorsal and anal fins are very dark. This is the situation in the southern part of the lake, but the distribution of both species in the remaining, northern,

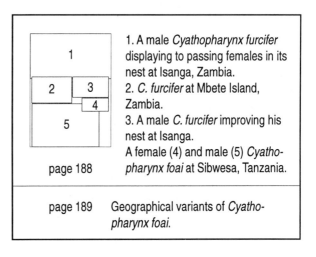

1. A male *Cyathopharynx furcifer* displaying to passing females in its nest at Isanga, Zambia.

2. *C. furcifer* at Mbete Island, Zambia.

3. A male *C. furcifer* improving his nest at Isanga.

A female (4) and male (5) *Cyathopharynx foai* at Sibwesa, Tanzania.

page 188

page 189   Geographical variants of *Cyathopharynx foai*.

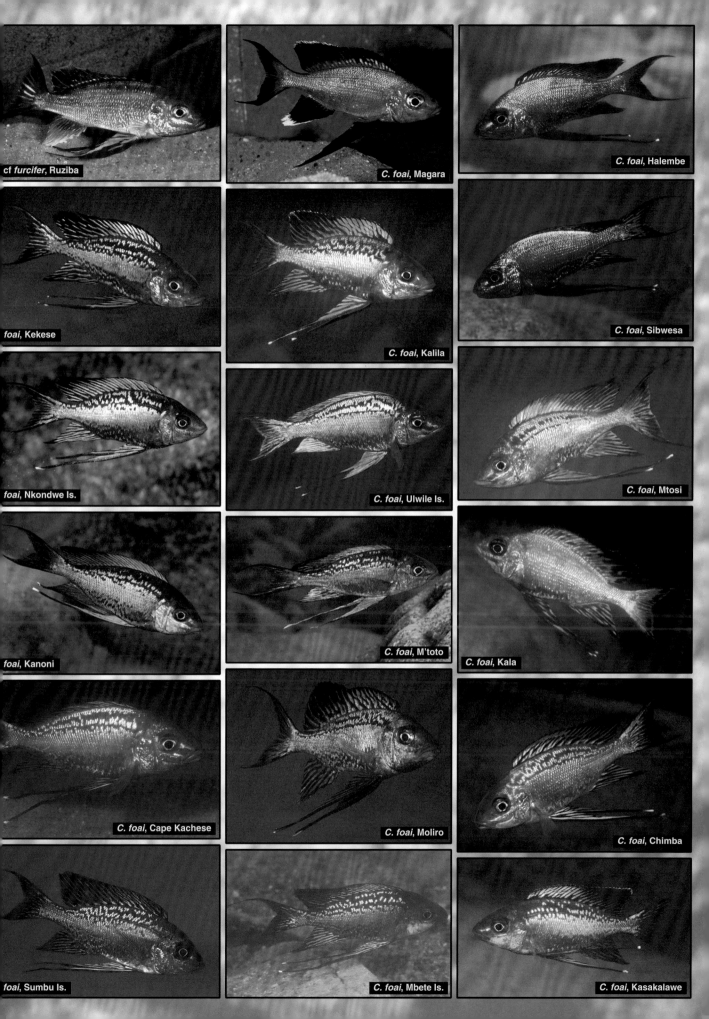

cf *furcifer*, Ruziba

C. foai, Magara

C. foai, Halembe

foai, Kekese

C. foai, Kalila

C. foai, Sibwesa

foai, Nkondwe Is.

C. foai, Ulwile Is.

C. foai, Mtosi

foai, Kanoni

C. foai, M'toto

C. foai, Kala

C. foai, Cape Kachese

C. foai, Moliro

C. foai, Chimba

foai, Sumbu Is.

C. foai, Mbete Is.

C. foai, Kasakalawe

part of the lake has not yet been resolved.

The male coloration of *C. furcifer* shows hardly any geographic variation in the southern half of the lake (it was found between Moliro Bay and Sibwesa), whereas along the same stretch of coastline I found four different colour forms of *C. foai*. I further found that in shallow areas with turbid water *C. furcifer* was either more abundant or the only *Cyathopharynx* present. On the other hand I found both species side by side at a depth of 22 metres at Mtosi, Tanzania. Males of both species construct sand-castle nests on top of rocks (see photos) but in the southern part of the lake I found only *C. furcifer* building such nests on the sand (at Kasanga and Kipili).

It thus seems that *C. furcifer* is a more flexible species, able to breed in various habitats. This would explain the observation that there are no obvious geographical differences between the known populations. *C. foai* on the other hand seems to be restricted more to the clearer water found in rockier habitats, which would explain the geographical variation observed in this species.

The situation in the northern half of the lake is still unclear. Herrmann (1994a) found two different *Cyathopharynx* species at Resha, Burundi (a shallow intermediate biotope): a dark blue male, apparently identical to the ones found at Rutunga, and one with marblings in the dorsal and a light-coloured anal fin. He further observed that males of both species constructed sand-castle nests on the sand. Which one of these two is *C. furcifer* (if this species is present at all in the northern half of the lake) can only be guessed, the assumption being that the lighter coloured fins are a characteristic of the northern *C. furcifer* as well.

The *Cyathopharynx* which has been shipped as the "Ruziba Furcifer" is very likely a different species from the one collected at Rutunga; it could in fact be the true *C. furcifer*. However, the identity of the gold-headed *Cyathopharynx* in Kigoma Bay, Tanzania, is far from clear. The golden patch on the head may indicate a relationship with the gold-headed *C. furcifer* from the southern regions of the lake, but the dark body and fins point to possible conspecificity with *C. foai*. It is clear that more research needs

to be done before these questions can be answered unambiguously.

More featherfins

The most common featherfin of the shallow intermediate habitat is *Aulonocranus dewindti*. It has a lake-wide distribution and is found mostly in habitats with coarse sand or gravel as substrate, at depths of less than five metres. Maximum total length is about 13 cm.

A. dewindti feeds on insect larvae and crustaceans and has, like the *Aulonocara* species of Lake Malawi, enlarged sensory pores on its head. The Malawi species use their sensitive sonar for detecting invertebrate prey that move around in the sand: they literally listen to the sand and hover motionless a few millimetres above the substrate. This very typical listening behaviour is not obvious in *Aulonocranus* and this species may therefore feed at night or at twilight, and use its enhanced sonar system to detect prey under subdued light conditions.

Territorial males form breeding colonies in which neighbouring males are two to three metres apart. They construct small nests with a diameter of about 30 cm, which are usually built against a small rock or between two adjacent rocks. *A. dewindti* nests can easily be told apart from those of other nest-builders because males carry pebbles from the surrounding area to decorate their spawning sites. Sometimes *A. dewindti* occurs in shallow and muddy bays without any rocks; here males construct sand-castle nests against water plants.

During the day females gather in groups and hover about 50 cm above the substrate. Males lead the females to their nests to lay their eggs. A female can produce up to 35 eggs per clutch.

The eggs are fertilized in the nest before collection by the female. Females do not nuzzle the male's ventral region or snap at the male's ventral fins; the male hovers a little above the female while she lays the eggs. The male may discharge his seminal fluid at this time or he may have done so in the nest before the female started to lay eggs. Mouthbrooding females again gather in nursery schools and stay in the shallow water where they release their offspring, either simultaneously or into schools of other juvenile fishes.

Ophthalmotilapia heterodonta

Between the Malagarasi River mouth in Tanzania and Nyanza Lac in Burundi, *Ophthalmotilapia heterodonta* shares the shallow intermediate habitat with *A. dewindti*. *A. dewindti* is predominantly found in the very shallow water whereas *O. heterodonta* is common at depths of three to five metres. *O. heterodonta* prefers phytoplankton and also scoops loose material from the biocover on rocks. Females congregate in schools in mid-water.

As far as is known, *O. heterodonta* is found only in the northeastern part of the lake, because *O. ventralis*, its closest relative, is reported to occur along the opposite shore at Luhanga (Kuwamura, 1986) and in other parts of the lake. In contrast to *O. ventralis*, male *O. heterodonta* build sand-scrape nests — pits which are dug alongside a rock or stone and appear crescent-shaped. These nests are similar to those of *A. dewindti*, i.e. sited against a rock, but no pebbles are used to decorate them. The crater wall is made of sand worked up from the centre of the nest and from fine sand taken from the surrounding area. Males have been seen stealing sand from each other's nests! Carrying a mouthful of sand, the male returns to his own nest, but does not deposit the sand directly on the rim. He first dives into the centre of the pit and shovels the sand from the centre, working it out of the nest so that it is deposited on the rim. There seems to be a continuous battle for fine sand; as soon as a male leaves his nest, other males take advantage and steal sand by the mouthful.

Male *O. heterodonta* are dark blue and clearly different from *A. dewindti* males. They can attain a maximum total length of about 14 cm. Spawning in *O. heterodonta* follows the pattern described for *O. ventralis* (page 42). Females deposit their eggs in the male's nest and are deceived by the egg-dummies on the male's ventral fins in order to fertilize the eggs inside the mouth.

Cunningtonia longiventralis

Cunningtonia longiventralis occupies a similar niche to *O. heterodonta* in the southern half of the lake. The breeding "mechanics" are similar in detail to those of *O. heterodonta*. Males build sand-scrape nests against rocks or between two adjacent rocks. Males lack the yellow lappets at the tips of the ventral fins but those tips are nevertheless a yellow colour.

C. longiventralis has thick fleshy lips with many teeth and feeds in a similar manner to *Petrochromis* species. It scrapes the biocover from rocks and stones and is therefore not in competition with *A. dewindti*, with which it shares the habitat. Females and juvenile males school together and may forage in pure rocky habitats as well. Often they are found mixed with juvenile *Ophthalmotilapia nasuta* in the rocky biotope.

The range of *C. longiventralis* encompasses the entire southern half of the lake and extends as far north as Kabimba along the Congolese

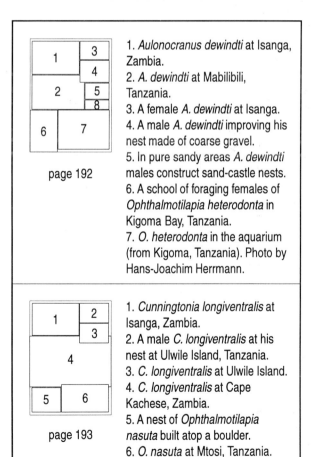

page 192

1. *Aulonocranus dewindti* at Isanga, Zambia.
2. *A. dewindti* at Mabilibili, Tanzania.
3. A female *A. dewindti* at Isanga.
4. A male *A. dewindti* improving his nest made of coarse gravel.
5. In pure sandy areas *A. dewindti* males construct sand-castle nests.
6. A school of foraging females of *Ophthalmotilapia heterodonta* in Kigoma Bay, Tanzania.
7. *O. heterodonta* in the aquarium (from Kigoma, Tanzania). Photo by Hans-Joachim Herrmann.

page 193

1. *Cunningtonia longiventralis* at Isanga, Zambia.
2. A male *C. longiventralis* at his nest at Ulwile Island, Tanzania.
3. *C. longiventralis* at Ulwile Island.
4. *C. longiventralis* at Cape Kachese, Zambia.
5. A nest of *Ophthalmotilapia nasuta* built atop a boulder.
6. *O. nasuta* at Mtosi, Tanzania.

shores of the lake (Poll, 1956). Its northernmost point along the eastern shore is Sibwesa (Poll, 1956). There are a few geographical variants known, as usual reflected in the male breeding coloration.

A featherfin with a nose for females

Ophthalmotilapia nasuta is a remarkable cichlid of the intermediate habitat, and it seems odd that it was not described until 1962 while species such as *Cyathopharynx furcifer* and *O. ventralis* were already furnished with a scientific name at the end of the 19th century. The fact that even females, which are rather common in the lake, have the "nose" characteristic of this species, serves further to emphasise the remarkably late discovery of this species.

As its scientific name suggests, the "nose" is the most characteristic feature of *O. nasuta*. Although the nose is evident in females, that of males, in particular breeding males, is much more obvious. The nose of *O. nasuta* does not have the same function as that of mammals and the term merely signifies the position of the swelling on the fish's head. It consists of a growth composed of skin and is not supported by a bony structure. Its function is not known, but since it is most developed in males, in particular territorial males, we may assume that it plays a role in breeding. The nose could, for instance, signal to conspecific males that the bearer is defending a spawning site, or be a sexual marker for females that are attracted to the males with the largest or darkest nose. The nose could also have a function in nest building.

The diet of *O. nasuta* consists of everything that drifts in the water column, even insects that have fallen into the water. In contrast to *C. longiventralis*, which is an algae-grazer, *O. nasuta* is a plankton-feeder that also feeds on microorganisms found just above muddy substrates. *O. nasuta* is often found in intermediate habitats with large boulders. Most individuals concentrate in foraging schools in mid-water about one to two metres above the bottom. Females are usually found in schools at depths of between three and ten metres; males are normally found a little deeper. Poll & Matthes (1962) found, after examining the catches of local fishermen, that there are three to four times as many females as males. This could be explained by the fact that males are normally found much closer to rocks and are thus in a better position to evade the fishermen's nets. Collectors of ornamental fishes have no problem collecting sufficient numbers of males and have not noticed any skewed ratio between the sexes.

O. nasuta is a maternal mouthbrooder. Males gather in breeding arenas and build sand-castle nests on top of large boulders which stand out from the surrounding rocks and must be very obvious to the females. Males take upon themselves the heavy burden of carrying enormous amounts, in comparison to the fish's size, of sand in their mouths, from the bottom to the top of the boulder. Sometimes the distance over which the sand is transported can be more than five metres! Interestingly the males compete for ownership of the highest rock, which automatically means more effort in the construction of a spawning site. The nests have a diameter of about 25-30 cm and can be 10 cm high. The weight of the sand is estimated at about two kilos. If you compare the weight of the male, at about 100 grams, to that of a human, then his efforts are equal to someone carrying 1.5 to 2 tonnes of books from the first to the 24th floor! We should consider ourselves lucky that our reproductive technique has evolved along a different route!

Territorial males are found two to three metres apart, and leks with 20 or more breeding males are common. The reason that males build their nests on top of rocks is possibly explained by the fact that they are then closer to the females or that their nests are more visible to them. Females forage in mid-water above the males' territories and it is simpler for the male to attract ripe females to his nest when it is visible. Males with nests at deeper areas have less chance of mating with females than those with nests close to them.

The spawning technique is similar to that of *O. ventralis* (see page 42). The eggs are fertilized inside the female's mouth when she draws in the seminal fluid released by the male while she tries to pick up the egg-dummies at the ends of his ventral fins. The female produces small batches of eggs, one to two dozen, which have a diameter of about 5 mm. Mouthbrooding females school together and feed during the last week of incubation, which lasts a total of about three weeks. The fry are released into schools of similarly-sized juveniles.

The holotype of *O. nasuta* was caught on the northwestern shore of the lake. The populations in that part of the lake are characterized by a very dark, almost black, body colour in the male. The black form of *O. nasuta* — underwater they appear to have a steel-blue colour — is found between Kalemie and Uvira. A second form, known among aquarists as the "Tiger Nasutus", occurs south of Kalemie along the western shore as far as Chimba in Zambia. Males of this form are characterised by a dirty yellow ground colour and a pattern of irregular spots and blotches on the head and anterior part of the body. The third form, the "Yellow Nasutus" or "Butterball Nasutus", has the widest range of all and is found in Zambian as well as in Tanzanian waters between Cape Kachese and Isonga and also between Halembe and Kigoma. Males of the populations at Cape Kachese and Kipili have the brightest yellow on the body.

Along the Mahali mountain range a differently coloured *O. nasuta* inhabits the intermediate habitat. It has some similarities to a form which has been exported under the name of "Nasuta Sela" and which purportedly originates from Moba in the Congo. Males have a dark yellow-bronze body colour and blue fins. The fifth form occurs in Burundi waters and is characterized by vertical bars on the body and a wide blue marginal band on the tail. Females of all known populations look similar: they have thin vertical bars on a ground colour which varies between yellowish and silvery.

Xenotilapia spilopterus

The genus *Xenotilapia* can be divided into two groups: one consisting of maternal mouthbrooders and the other of species which form pairs during the breeding period and employ a biparental mouthbrooding technique. *X.*

spilopterus belongs to the group of biparental mouthbrooders.

When not breeding, *X. spilopterus* lives in groups, sometimes in large schools over the sand of the intermediate habitat. It forages by scooping up the sandy substrate or sediment and sifting it through its gills. Its main interest seems to be insect larvae. Rarely large schools are found in mid-water where they appear to feed on zooplankton, but normally they stay near the bottom.

With the approach of the breeding season — at present it is not known whether there is a regular annual breeding season, or whether breeding is triggered by the effect of some external stimulus on the members of the school — the foraging school splits up into pairs. Each pair stays in a territory with a diameter of about two metres, which is defended mainly against conspecifics.

The pair bond is established by repeated courtship displays by the male as well as by the female. In the artificial environment of the aquarium it seems that the bond between the pair is strengthened by the continuous protection of the territory. If only one pair is kept in the tank there is a risk that the male and female will quarrel and that one of them, not always the female, might end up cowering in a corner. Even when the pair is not yet brooding, male and female stay together in a relatively small area.

It is difficult to predict an imminent spawning, but mutual courtship increases noticeably, sometimes days before the actual spawning. A slight change in the colour pattern occurs as well; the female as well as the male acquires black markings in the upper and lower parts of the iris. These markings, together with the pigment in the eye, form a black vertical bar across the eyeball. I don't know if this is a sign of

page 196	*Ophthalmotilapia nasuta* has a lake-wide distribution and occurs in many different geographical variants.
page 197	Some variants of *Xenotilapia spilopterus*.

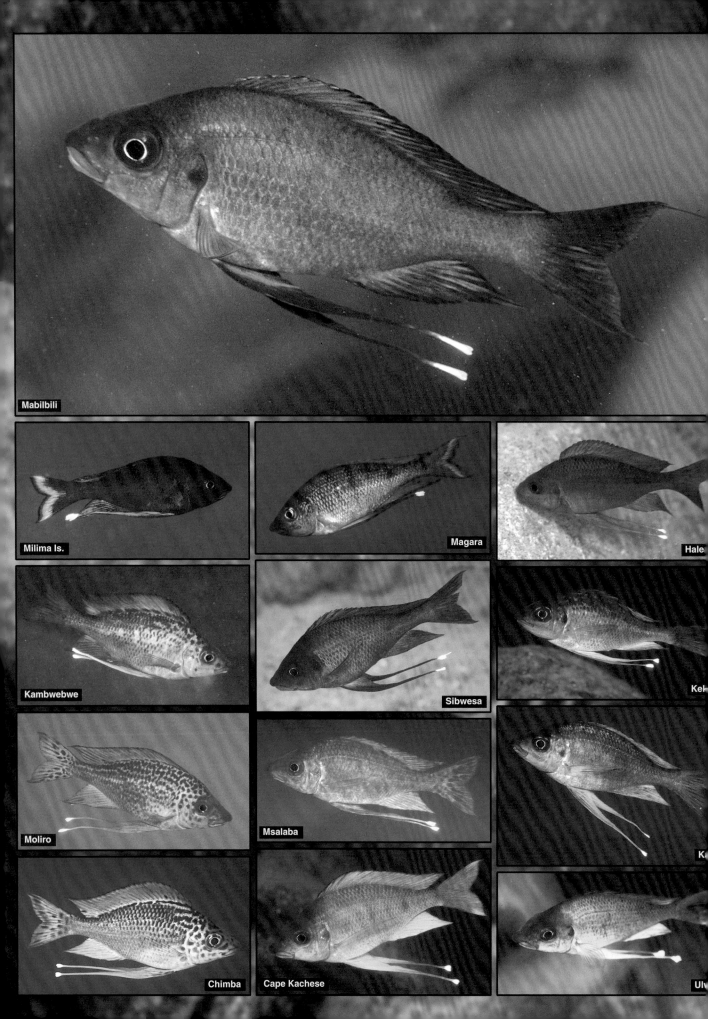

Mabilbili

Milima Is.

Magara

Hale

Kambwebwe

Sibwesa

Kel

Moliro

Msalaba

Ka

Chimba

Cape Kachese

Ulv

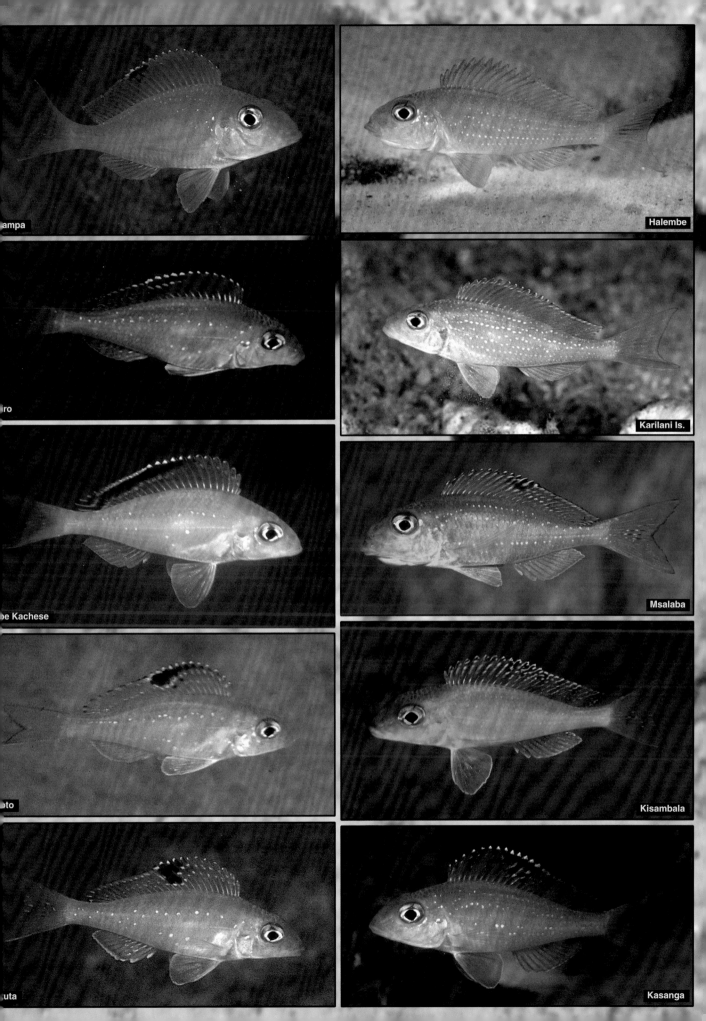

ampa

Halembe

ro

Karilani Is.

be Kachese

Msalaba

to

Kisambala

uta

Kasanga

readiness to spawn, but it is seen mostly when the pair is engaged in spawning or shortly before the act.

Spawning takes place inside the pair's territory but there is no specific site or nest where the eggs are deposited. In fact, the deposition site can change during spawning. In the maternal mouthbrooder group of *Xenotilapia*, male and female circle around each other before the eggs are laid, but *X. spilopterus* starts spawning when the female suddenly deposits some eggs on the substrate. I have never noticed a signal from the male for the female to start. The male usually waits behind the egg-laying female, about 3 cm above the substrate, until she vacates the site and leaves the eggs to be fertilized by him. While the male positions his vent over the eggs and discharges seminal fluid, the female turns around and waits until the male moves on. Then she picks up the eggs. The female will not lay any eggs if the male is not behind her. While he is chasing intruders from the territory she may remain at the site, but she may also join the male in the territorial defence. After a short interruption the female again lowers her body to the substrate and waits for the male to take position behind her before she lays a new batch of eggs.

Spawns can be as large as 40 eggs, but much seems to depend on the buccal capacity of the female. If the female's mouth is getting full she has to shuffle the eggs around before she can pick up new ones. Usually the male has his head near the eggs when the female collects them and sometimes picks up some eggs himself. Once I noticed that he picked up two eggs, which the female apparently didn't collect quickly enough, and swam away while chewing on them. Then the pair went through another cycle, and when the female went to pick up the new batch of eggs, the male spat out the two eggs in front of her, and she immediately picked them up. During the next round, the female's mouth was probably too full, and it took so long before she arranged the eggs that were already inside her mouth that the male again took some eggs, but this time he ate them. The female didn't produce any more eggs after this incident.

Shortly after spawning the pair reduce the size of their territory, but the smaller area is defended more vigorously than usual. For the first nine to twelve days the female broods the embryos and refrains from feeding. After this period the larvae are transferred to the male's mouth. I have never witnessed the procedure in its entirety, but it is likely that, following some sort of preliminary ritual, the female spits all the larvae in front of the male. The male retrieves them quickly and continues the brooding for another six to ten days while the female stays by his side.

When brooding is complete, the male releases the fry in the territory where they are protected by both parents. For the first few days the fry take refuge inside the male's mouth but most of the time they sit on the substrate. They sometimes wander off into other territories where they are protected by the resident pair. The fry measure about 15 mm at the time of release and it takes them about two years to reach the adult size of approximately 10 cm.

The biparental mouthbrooding technique is not more advanced than the maternal mouthbrooding procedure. While mouthbrooding pairs have to devote three to four weeks to taking care of one brood, males of a maternal mouthbrooder can fertilize eggs of many females in the same period. In theory, this should not make a difference as long as there are as many females as males. The drawback of biparental mouthbrooding is the fact that a territory has to be maintained by the pair. In maternal mouthbrooders only the strongest males occupy a territory, and the mouthbrooding females gather in nursery schools. Where population density is high, as is the case in Lake Tanganyika, this offers a distinct advantage. Instead of having many territories spread over a large area, a species can propagate in a relatively small region of the biotope. Only the males have to worry about a territory and since there are fewer needed, any losses during the selection process probably enhance the viability of the species. The advantage of biparental mouthbrooding is that the offspring are protected by both parents, which may be necessary in rocky environments where predators can hide very close to the fry.

The stimulus for the female of the biparental mouthbrooder to transfer the larvae to the male's mouth may be hunger. During the 9 to 12 days incubation she does not eat at all. Females of a maternal mouthbrooder endure a longer period without food, but some species eat small morsels in the second half of the incubation period. Such mouthbrooding females also spend less energy than those of a biparental mouthbrooder which also have to defend a territory.

The holotype of *X. spilopterus* was caught at Nkumbula Island (Crocodile Island) near Mpulungu in Zambia, but its range is much larger. In the Congo I have found it south of Kapampa, but it may have a wider distribution in the northern part. It also occurs in Zambian and Tanzanian waters, but has not been found in Burundi.

Several geographical variants of *X. spilopterus* are known and some have been exported as aquarium fishes. Most variants have a colourless dorsal fin with some black markings on the edge. The variant found between Kigoma and Kasoje has a dorsal fin with tiny, coloured spots, but lacks the blotchy markings found in all other known populations; it is known as the "Pearly Spot Spilopterus". The variant that inhabits the rocky shores between Mabilibili and Kasoje in Tanzania has an attractive yellow coloration in addition to the black markings in the dorsal fin.

Kuwamura (1986a) reports biparental mouthbrooding in *X. longispinis* on the northwestern coast of the lake. The data provided and a photograph published in 1997 (p 66) suggest that there has probably been a misidentification. *X. longispinis* attains an adult size of about 16 cm while the nine females examined by Kuwamura have an average total length of about 8.5 cm. Judging from the photograph it appears that *X. spilopterus* or a species with a close resemblance also occurs in that part of the lake. In 1987 the same author recorded *X. longispinis* from Miyako on the central east coast of Tanzania. The variant of *X. spilopterus* found here, the Pearly Spot, has also been erroneously identified (in Brichard, 1989: 432 and 433) as *X. longispinis*. The Pearly Spot may be an undescribed species, but its behaviour and morphology are so close to *X. spilopterus* that a specific distinction seems unnecessary.

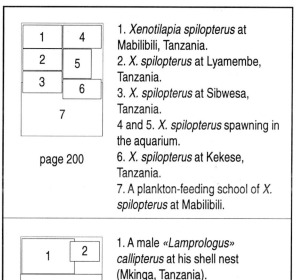

1	4
2	5
3	6
7	

page 200

1. *Xenotilapia spilopterus* at Mabilibili, Tanzania.
2. *X. spilopterus* at Lyamembe, Tanzania.
3. *X. spilopterus* at Sibwesa, Tanzania.
4 and 5. *X. spilopterus* spawning in the aquarium.
6. *X. spilopterus* at Kekese, Tanzania.
7. A plankton-feeding school of *X. spilopterus* at Mabilibili.

1	2
3	
	4
6	5

page 201

1. A male «*Lamprologus*» *callipterus* at his shell nest (Mkinga, Tanzania).
2. *Lepidiolamprologus hecqui* at Mbete Island, Zambia.
3. An empty shell with a pair «*Lamprologus*» *brevis* at Chimba, Zambia.
4. «*L.*» *brevis* (left) and *Neolamprologus ocellatus* (right) at Katete, Zambia.
5. A female «*L.*» *ocellatus* with young at Isanga, Zambia.
6. A shell bed with *Telmatochromis brichardi* at Kerenge Island, Tanzania.

Empty shells

Large empty shells, mainly of the snail *Neothauma tanganyicense*, are found on sandy or muddy bottoms but rarely among the rocks. Quite frequently, in particular in regions deeper than 10 metres, large accumulations of empty and partly crushed shells provide a calciferous "rocky" habitat for a group of small, "cave-brooding" cichlids.

Three different shell-biotopes can be distinguished. The large accumulations understandably offer the greatest living space but usually just a few species dominate the communities in such habitats. In shell beds of this type, which can be more than several metres thick and several kilometres long, most top-layer shells are "glued" together by calciferous deposits which can completely surround the shell, closing it off, and render it useless as a cichlid home.

The second type of shell biotope consists of loosely arranged shells grouped together in shallow habitats and usually found near rocks. I have found such groups mostly in the upper 15 metres. It is possible that all shell groups of this type are made by cichlids (see below). The third type consists of single empty shells lying scattered on the sandy or muddy lake floor. Such a pattern is rather common on many sandy shores, with shells found in shallow as well as in very deep water.

Surrogate caves

Over millions of years the fauna of Lake Tanganyika has developed, independently from those of other drainages, into the rich and varied aquatic communities we see today. About 200 cichlid species are known to inhabit the lake, most of them specialized for a particular biotope and a particular ecological niche. Such a niche can be defined as the environmental requirement of a species, or its compatibility with its environment, in terms of characters such as the topography of the biotope, the type of food, the depth, the type of spawning site, etc. The great number of different species found at many localities suggests that many different niches have been exploited.

As regards brood-care technique, the cichlids of Lake Tanganyika can be divided into two main groups: the substrate brooders and the mouthbrooders. Almost all the substrate brooders affix their eggs to a hard substrate and are hence found mostly in rocky habitats; moreover the majority of these substrate brooders are small species which seek shelter in the holes and caves of the rocky habitat. Except for a few large piscivorous species, all the substrate brooders are cave brooders. Not only for breeding purposes, but also for their own protection and that of their offspring, these cichlids need the security of the caves and crevices the environment provides.

Generally speaking, the shell of a freshwater snail degrades (in fact dissolves) rather rapidly after the animal has died. However, in Lake Tanganyika the high pH (alkalinity) of the water prevents rapid degradation of the empty shells. On the contrary, more calcite is deposited onto those shells that are exposed to lake water, making them thick and hard. Buried shells remain virtually unaltered for very long periods of time — thousands of years (Cohen et al. 1993)! — and only mechanical degradation of shells occurs. This means that these empty shells provide long-term caves for any fishes able to fit into them. However, although there are many different species of snails and bivalves in the lake (most of them endemic), the shells of only a few species are large enough to offer shelter to cichlids, or, to look at it a different way, only a small number of cichlid species are small enough to take refuge inside the available shells.

For these small cichlids empty shells are the equivalent of caves among the rocks, and so the shell-brooding cichlids are cave-brooders like most other substrate brooders.

Tailor-made cichlids

The most important prerequisite for a cichlid wishing to benefit from the protection offered by an empty shell, is a size small enough to fit into it. As mentioned before, empty shells are found on the open sand, an area where small cichlids are very vulnerable prey. In only a few species are both adults small enough to shelter into shells. There are, however, also species in which only the females are small enough to fit inside the shell, whereas the males, usually considerably larger than the females, have to find shelter at another location. Then again, a number of species have adopted the empty shells as a nursery for their young but are themselves too large to fit inside. Most often one finds such species at the edge of rocky habitats where the parents can find refuge among the rocks, but where the rocky habitat does not provide adequate nesting sites or where population densities of cave-brooders are too great (so that some individuals have to search for alternative breeding sites). Species that have been found using empty shells as a nesting site, but not for their own shelter, are *Lepidiolamprologus attenuatus*, *L. pleuromaculatus*, *«Lamprologus» caudopunctatus*, *N. mondabu*, *N. cunningtoni*, *Altolamprologus fasciatus*, *A. compressiceps*, and *Telmatochromis temporalis* (Sato & Gashagaza, 1997).

Most of these species are also found on the large shell beds, but here they lack the protection of rocks nearby. Some of them have developed miniature forms which are able to fit inside the empty shells. One such species, well-known in the aquarium hobby, is the so-called "Shell Compressiceps", which lives in and around empty *Neothauma* shells. This appears not to be a distinct species but simply a miniature form of *A. compressiceps* which has developed at several places around the lake. The first shell-brooding *A. compressiceps* were caught near Sumbu Island in Zambia, but similar forms are found near Utinta in Tanzania, Rumonge in Burundi (Gashagaza *et al.*, 1995), and Katibili (as *Altolamprologus cf. calvus*) in the Congo (Büscher, 1998). However, when these shell-brooding *A. compressiceps* are kept in the aquarium it appears that they outgrow their shells and attain sizes more common for *A. compressiceps*. Gashagaza *et al.* (1995) have recorded two more species which have developed miniaturized forms on shell beds: *Neolamprologus mondabu* and *«L.» callipterus*. The latter is a shell-brooding cichlid in which usually only the female fits inside empty shells. At Rumonge the authors found a form in which the male too was small enough to fit the shells.

Of great interest was the discovery by Büscher (1998) of a miniature form of *«Lamprologus» speciosus*, the normal form of which is a shell-brooding species. In most places this species inhabits empty *Neothauma* shells, but at Kasenga, Congo, the entire population used the noticeably smaller *Lavigeria* shells. I have found miniaturised forms of another shell-brooder, *«L.» brevis*, at Karilani Island, Tanzania, but the most interesting part of Büscher's discovery was that when he placed the miniature *«L.» speciosus*

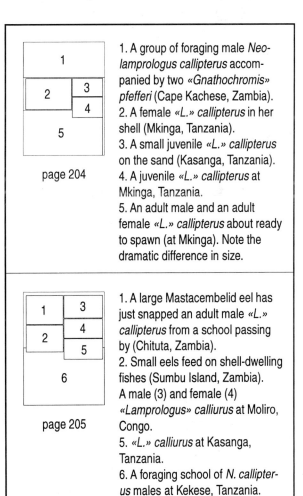

page 204

1. A group of foraging male *Neolamprologus callipterus* accompanied by two *«Gnathochromis» pfefferi* (Cape Kachese, Zambia).
2. A female *«L.» callipterus* in her shell (Mkinga, Tanzania).
3. A small juvenile *«L.» callipterus* on the sand (Kasanga, Tanzania).
4. A juvenile *«L.» callipterus* at Mkinga, Tanzania.
5. An adult male and an adult female *«L.» callipterus* about ready to spawn (at Mkinga). Note the dramatic difference in size.

page 205

1. A large Mastacembelid eel has just snapped an adult male *«L.» callipterus* from a school passing by (Chituta, Zambia).
2. Small eels feed on shell-dwelling fishes (Sumbu Island, Zambia). A male (3) and female (4) *«Lamprologus» calliurus* at Moliro, Congo.
5. *«L.» calliurus* at Kasanga, Tanzania.
6. A foraging school of *N. callipterus* males at Kekese, Tanzania.

203

in an aquarium with large shells they attained the same length as conspecific individuals that normally live in such shells.

So it appears that the size of some species is controlled by their environment rather than their genes. These dwarfed forms are fully mature and breed inside the shells. They may be exposed to a large group of predators that cull the population of shell-brooders, leaving only those able to hide inside shells. Dwarfed forms usually have a lighter coloration which blends in better with the light background formed by shells and sand (giving them additional protection from predators).

Two shell-brooding species are known in which only the females live inside the shell while the males, usually grouped in large packs, roam through the intermediate habitat where they find protection in the school. Single males need the protection of the caves among the rocks. This may be the reason why females of these two species, «*Lamprologus*» *callipterus* and «*L.*» *calliurus*, live in shells that are close to rocks or in intermediate habitats.

Shells and harems

Males of «*L.*» *callipterus* can attain a total length of about 15 cm and are strong enough to carry empty shells in their mouths. While many mouthbrooding cichlids in the lake build nests or spawning sites of sand, the cave-brooder «*L.*» *callipterus* carries empty shells from the surrounding area to its nest site. All that males have to offer to females is a collection of empty shells near the rocky habitat. In order to control sites of this type the shells — some nests contain more than 100 shells — are placed close together on a patch of substrate with a diameter of about 40 cm. Neighbouring males are always on the prowl, hoping to steal some shells from other nests, and it is not uncommon to see males carrying shells between nesting sites. Females, which attain a maximum total length of about 6 cm, usually hide inside shells and when a female is ready to spawn she looks for an attractive nest. Males can hold harems of more than 20 females at a time.

The female deposits her eggs inside the shell but there is no possibility of the male entering or even seeing what is going on. The female must therefore signal to the male that she is ready to spawn and that she expects him to fertilize the eggs. Once you have seen this in the wild you will appreciate the enormous difference in size between the male and female (see photo page 204) and the way the female courts the male by swimming agitatedly around his head. Once the female has got the male's attention she disappears inside her shell and starts laying eggs. The male now takes up position outside, with his vent above the opening to the shell, and discharges his milt which is then fanned inside the shell by him and also sucked inside the shell when the female backs out. The eggs are fertilized inside the shell and the female takes care of them. When the fry swim free they stay around the nest for a while, but the female abandons them and searches for a new shell, outside the male's territory, in which to recover.

When a male steals shells which contain fry from competitors' nests, he will eat these fry before the shell is added to his own nest (Gashagaza et al., 1995). This instinctive measure prevents the male from guarding offspring which do not carry his genes. This may sound simple enough, but in fact this instinctive behaviour represents an incredible piece of evolution during the course of the lake's history, especially if we take into consideration the fact that a number of other species are tolerated or completely ignored inside the male's territory.

When «*L.*» *callipterus* juveniles attain a total length of about 3 cm they leave the shell area and are found, usually in small groups, not only among the rocks of the intermediate zone but also in pure rocky habitats. At a total length of about 7 cm they join other groups of similar-sized «*L.*» *callipterus* and form large foraging schools which can contain more than 100 individuals. They feed mainly on shrimps, and since this type of prey seems to be difficult to locate in the habitat — shrimps can be detected only when they are disturbed and flee away — their concerted effort proves beneficial to each individual (Yuma & Kondo, 1997). «*L.*»

callipterus exposes shrimps by blowing away (by squirting water from the mouth) the silt and sediment covering the substrate. Large foraging schools are rather common, and apart from eating shrimps they also feed on the fry of substrate-brooding cichlids whenever they happen on them. Large schools also uncover large quantities of prey, and other species, such as *«Gnathochromis» pfefferi* and sub-adult *Lepidiolamprologus elongatus*, may join *callipterus* schools in order to profit from their activity.

«L.» callipterus has a lake-wide range and is very common in most areas, and in this way provides other cichlids, which are allowed access to the shells, with additional brooding substrate near rocky areas.

A second shell-brooding species, *«L.» calliurus*, likewise consists of small females and large males, but in contrast to *«L.» callipterus* these males are not strong enough to carry empty shells and thus depend on existing shell distributions.

Males of *«L.» calliurus*, which can attain a total length of about 10 cm, are sometimes also found in large aggregations while feeding on plankton. Males temporarily defend groups of shells and allow females to settle in these. Their harems are usually no larger than five females, and several males may have adjoining harems in larger shell beds. Adult females are much smaller, maximum total length approximately 4 cm, and live in shells for their entire lives. Although I have observed males defending a few females against competing males, it appears that males do not hold territories for a long time. It is possible that males fertilize a few spawns and then abandon the females to care for the offspring, because groups of females are more common than territorial males defending them.

In the past *«L.» calliurus* has been confused with *«L.» brevis*, a much smaller species with a rather different breeding behaviour (see page 214). The difference between the males of these two species is simple: *«L.» calliurus* males have lyre-shaped tails while those of *«L.» brevis* males are rounded. Female *«L.» calliurus* are a much lighter colour and their tails are straight; small females are, however, difficult to tell apart from those of *«L.» brevis*. Both species are found throughout the lake.

«Lamprologus» ocellatus and its look-alikes

Apart from *«L.» callipterus*, all obligatory shell-brooding cichlids are too small to transport shells and must accommodate their requirements to the existing distribution of shells. At many localities empty shells are widely scattered and a frequency of one to five shells per square metre seems to be normal. In such areas you will not find *«L.» callipterus* or *«L.» calliurus*, or a number of shell-brooders that appear to be specialized in living in areas with a much higher shell density. A few species are, however, found mainly in these low-density shell areas, and the most common of these is *«Lamprologus» ocellatus*.

«L.» ocellatus has a very large, almost lake-wide distribution, but is not found between Kalemie and Moliro along the Congolese shore, where its particular niche seems to be occupied

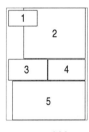

1. A male *«Lamprologus» ocellatus* in front of his shell (Katete, Zambia).
2. A female *«L.» ocellatus* guarding her offspring (Isanga, Zambia).
3. *«L.» stappersi* in the aquarium (photo By Mark Smith).
4. *«L.» speciosus* in the aquarium (photo by Hans Herrmann).
5. A female *«L.» ocellatus* protecting her shell (Isanga, Zambia).

page 208

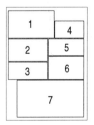

1. A male and female (adult!) *«L.» ornatipinnis* in front of the female's shell in Kigoma Bay, Tanzania (photo by Horst Dieckhoff).
2. A male and 3. a female *«L.» ornatipinnis* (Kigoma, Tanzania; photos by Horst Dieckhoff).
4. Female and 5. male *«L.» sp.* "ornatipinnis zambia" in the aquarium.
6. A male *«L.» sp.* "ornatipinnis zambia" at his shell (Cape Kachese, Zambia).
7. *«L.» sp.* "ornatipinnis zambia" at Cape Chaitika, Zambia.

page 209

by two different species: «L.» stappersi (better known among aquarists as «L.» meleagris) and «L.» speciosus. «L.» stappersi occurs in shallower water than «L.» speciosus but both species are found side by side (Büscher, 1991a, 1998).

When Büscher (1991) described «L.» meleagris he pointed out that it differed very little from «L.» stappersi, but that the type locality of the latter species was about 100 km further north of that of «L.» meleagris, which, at that time, was known only from the area near Bwassa (65 km south of Moba). Later he also recorded «L.» meleagris from Kalemie, where it was found in the river mouth (Büscher, 1998). The holotype of «L.» stappersi was collected near Mpala, in or near a river. This locality lies between Moba and Kalemie, and it seems very unlikely that such similar forms, both found on muddy bottoms near river mouths, are not conspecific. Although it is agreed that «L.» stappersi was initially poorly described, Büscher has re-described this species, and there is no doubt that «L.» meleagris is synonymous with «L.» stappersi, originally described by Pellegrin in 1927.

«L.» speciosus may in the future experience a similar fate. In his description of «L.» speciosus Büscher failed to compare his new species with «L.» wauthioni because at that time (1991) the latter was not included in Lamprologus. However, the only reason «L.» wauthioni had not been included in Lamprologus was because of its pointed ventral fins; all species in Lamprologus have rounded ventrals. Following Stiassny's recent (1997) revision, there are no Lamprologus species in the lake (see page 22), but the fact remains that «L.» speciosus does not have pointed ventral fins, at least not in the populations known. In his description of «L.» wauthioni Poll (1949) mentions the characteristic, round, light spots in the tail and the posterior parts of the anal and dorsal fins. These spots were still evident when I photographed the holotype almost 40 years later. «L.» wauthioni hails from the same area as «L.» stappersi and «L.» speciosus, and if one compares its characteristic markings with those of preserved «L.» speciosus (Büscher, 1998: 55), then one wonders whether pointed fins could be part of some geographical variation within a species. However, until such time

as «L.» speciosus with pointed ventrals are discovered, we must continue to regard «L.» speciosus and «L.» wauthioni as separate species.

Both males and females of the aforementioned ocellatus-like species occupy empty shells, and, as far as is known, males hold harems of two to five females. It is known that males of «L.» ocellatus defend territories in which several empty shells are located. One of these shells is used as the male's home and the others are completely covered with sand (Walter & Trillmich, 1994). In fact the "home" shell is also buried into the sand, but the opening is left free. By burying the spare shells the male restricts their usage to females he wants to mate with (see below) and prevents them from being occupied by competing males or other species. He may also bury them to protect them from being carried away by «L.» callipterus, although the latter species is most often found near rocky habitats while «L.» ocellatus is at home on wide open sandy bottoms.

It is easier to bury shells in fine-grained sand and mud than in coarse sand, and this is the reason why ocellatus-like shell-brooders occur mainly on sand/mud bottoms near river mouths or in muddy bays. Moreover, another way of avoiding predators, when empty shells are not immediately available, is to dive into the bottom and remain motionless until the threat has (hopefully) disappeared. Even small juveniles practice evasive dives of this type.

These small shell-brooders are greedy eaters which feed mainly on insect larvae, e.g. those of mosquitos and similar insects.

A female that are interested in settling in the male's territory must first persuade her intended mate to uncover one of his buried shells for her. A compromise is necessary between the number of females a male can afford in his territory and the distance between individual shells. Several females may settle in the male's territory, but, as females can be aggressive among themselves, their shells must be well-spaced. Too close together means too many squabbles between the females and fewer offspring per female, and too far apart means that he cannot control his females and may lose some to other males. The most common distance between the females is about 80 to 100 cm.

Before the actual spawning the female has again to win the male's attention, because the eggs are deposited inside her shell. She swims up to the male, bends her body with the vulnerable belly region towards him, and quivers. After a few seconds of quivering she proceeds to her shell where she waits for the male, who usually follows immediately behind her. Still quivering she enters the shell and starts laying eggs. After a few eggs have been deposited she backs out of the shell and now the male enters to fertilize them. In the aquarium I once observed two males fertilising the spawn of a single female (Konings, 1980), but whether this occurs in the wild is doubtful.

Herrmann (1995) reports that in «L.» stappersi the male assumes a stiff and awkward pose at the shell's entrance while the female deposits the eggs inside, and expels seminal fluid towards the female. This too may not be an accurate reflection of what happens in nature: in the case reported the male may have been too large to fit into the shell and had resorted to fertilizing the eggs from the outside, or this could be normal procedure when the female's shell is too small for the male.

The fry are free-swimming in about 10 days and are cared for by the female. Juveniles can remain in the male's territory for more than two months and find shelter in any of the available shells, including that of the male. On the other hand fry are sometimes cannibalised by the male and by other, wandering, conspecifics. For this reason the female is very aggressive in the defence of her offspring, in particular during the first few days after they leave the shell.

«Lamprologus» ornatipinnis

There are at least two species of «L.» ornatipinnis-like cichlids present in the lake, but the fish that is usually associated with this name in the aquarium hobby is not the nominal species. «L.» ornatipinnis was described by Poll (1949) from specimens (78 types and 55 additional specimens) collected all over the lake. The holotype was caught south of M'toto near the barrage of Moba. Remarkably Poll was unable

to find a female among the 133 specimens. The species appears to have a lake-wide distribution although it has not yet been reported from Zambian waters. The probable reason is that «L.» ornatipinnis does not occur in Zambia, where another, similar, species inhabits the muddy bottoms. This species is the one that has been around in the aquarium hobby for at least 15 years and now appears to be the one without a scientific name. It is now therefore named «L.» sp. "ornatipinnis zambia". The maximum total length of this shell-brooder is about 6 cm for males and 4.5 cm for females.

I have seen "Ornatipinnis Zambia" at Cape Kachese, Cape Chaitika, Mbete Island, and at Isanga (all localities in Zambia), and although it was never abundant, it seems to be a common species. Its behaviour is very similar to that of «L.» ocellatus, and both male and female have their own shell, usually Neothauma. It is not known whether males — which, like ocellatus-like species, have harems — hide shells in the sand until they are required for interested mates. An interesting feature is that «L.»

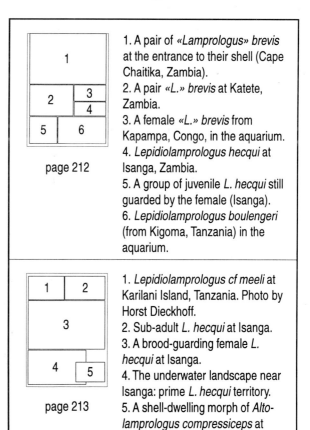

1. A pair of «Lamprologus» brevis at the entrance to their shell (Cape Chaitika, Zambia).
2. A pair «L.» brevis at Katete, Zambia.
3. A female «L.» brevis from Kapampa, Congo, in the aquarium.
4. Lepidiolamprologus hecqui at Isanga, Zambia.
5. A group of juvenile L. hecqui still guarded by the female (Isanga).
6. Lepidiolamprologus boulengeri (from Kigoma, Tanzania) in the aquarium.

page 212

1. Lepidiolamprologus cf meeli at Karilani Island, Tanzania. Photo by Horst Dieckhoff.
2. Sub-adult L. hecqui at Isanga.
3. A brood-guarding female L. hecqui at Isanga.
4. The underwater landscape near Isanga: prime L. hecqui territory.
5. A shell-dwelling morph of Altolamprologus compressiceps at Sumbu. Photo by Horst Dieckhoff.

page 213

ocellatus buries its shell with the opening commonly at a right angle to the bottom while «*L.*» sp. "ornatipinnis zambia" completely buries the shell with the opening almost flush with the bottom (see photos pages 208 and 209). «*L.*» sp. "ornatipinnis zambia" feeds on insect larvae and other bottom-dwelling invertebrates and not so much on plankton like «*L.*» *ornatipinnis*.

«*L.*» *ornatipinnis* — previously referred to as *Lamprologus* sp. aff. *ornatipinnis* (Konings, 1988) — was first seen in its natural habitat by Dieckhoff in Kigoma Bay. He found that females were remarkably smaller than males (3 cm vs. 8-9 cm) and that several females could be found in the male's territory. The females occupy small shells, probably those of *Paramelania* and *Lavigeria* snails, and in this respect differ little from the Zambian species. Males, however, are too large to fit inside a shell and patrol the territory; they flee over the bottom when danger threatens.

Male «*L.*» *ornatipinnis* feed on *Cyclops*-like crustaceans that hover a few centimetres above the muddy substrate (Poll, 1956). Females have not been examined, and because they have a different lifestyle, hiding in shells, may actually feed on something different.

Spawning takes place inside the female's shell, and the eggs are fertilized by the male discharging his milt in front of the entrance to the shell and fanning it towards the eggs. Moreover, when the female backs out of the shell there is an influx of water (containing semen) into the shell.

Büscher (1998) recorded «*L.*» *ornatipinnis* or a similar species — designated *L.* sp. *aff. ornatipinnis* — from Tembwe (Tembwe Deux) in the Congo, about 40 km south of the type locality of «*L.*» *ornatipinnis*. Like Poll he found many males, but also two tiny (but adult) females inside small shells. The males he was able to catch had a maximum total length of 9 cm, whereas the maximum size of the females was 3.3 cm.

A morphological difference between «*L.*» *ornatipinnis* and the undescribed species from Zambia exists in the shallower and more elongate body of «*L.*» *ornatipinnis*, and in addition the stripes in the dorsal fin are more numerous than in Ornatipinnis Zambia.

The most specialized shell-brooder

We have seen that there are at least three species of shell-brooders, «*L.*» *callipterus*, «*L.*» *calliurus*, and «*L.*» *ornatipinnis*, in which only the female hides inside the shell and the male needs to find other shelter when danger threatens. Another group of shell-brooders, «*L.*» sp. "ornatipinnis zambia", «*L.*» *ocellatus* and allied species, have small males able to hide in shells — each in its own shell. There is, however, a species with a lake-wide distribution in which both male and female hide in the same shell: «*Lamprologus*» *brevis*. Whereas the members of the first two groups are harem breeders, «*L.*» *brevis* is usually monogamous. This is probably a logical consequence of its lifestyle; a single shell, usually *Neothauma*, is sufficient for a breeding pair. «*L.*» *brevis* is thus able to live and thrive in areas with an extremely low density of empty shells and sometimes inhabits regions where there are no other shell-brooders.

«*Lamprologus*» *brevis* is the only cichlid known thus far in which both male and female reside in the same shell. When threatened, the female, the smaller of the two (maximum total length about 4 cm), enters the shell first, followed by the male, who has a maximum size of about 5 cm. The shell is usually not covered or hidden in the sand although shells abandoned by other shell-brooders are accepted as shelter. «*L.*» *brevis* is a plankton-feeder and pairs are most often found hovering above their shells picking at zooplankton drifting past their home.

Spawning takes place in the same shell, and when the fry swim free the entire family hides in the single shell. The fry leave the parental shell at an early stage, about one week after becoming mobile, and wander about over the lake floor in search of shells of their own. The cryptic coloration of the fry blends well with that of the substrate, an essential camouflage as the fry are usually unable able to find empty shells to occupy because of the territorial demands of adults. It appears that only adults are able to occupy shells and that juveniles need to rely on their cryptic coloration and careful movements in order to survive. «*L.*» *brevis* is another example of the extent to which cichlids have adapted to each and every possible niche available in Lake Tanganyika.

Other shell-brooders

In areas with a low density of empty shells there are other shell-brooders which seem to have derived from a different group of cichlids. Of the shell cichlids discussed above probably only «L.» callipterus is derived from a different ancestor — the others seem closely related. There are also several species of the genus Lepidiolamprologus that use empty snail shells as brood caves, and we have seen earlier (see page 182) that L. attenuatus and a closely-related species use an empty shell as a nursery for their offspring.

Lepidiolamprologus boulengeri and L. meeli (often confused with L. hecqui) dig a relatively large pit in which one or more shells provide protection for their offspring. These species appear to be monogamous and use the empty shell(s) mainly as spawning receptacles. Male L. boulengeri and L. meeli attain a total length of about 7 cm whereas females remain about 5 cm in length. When disturbed it is usually the female that takes refuge inside the shell while the male flees over the sand. The fry live on the bottom and rely on their cryptic coloration for protection from predators.

L. boulengeri (L. kiritvaithai is a junior synonym) occurs in the northeastern part of the lake, between Nyanza Lac in Burundi and the Malagarasi river mouth in Tanzania. The holotype of L. meeli was caught south of Kalemie, near Katibili on the west side of the lake, but a population of a very similar form on the opposite side, at Karilani Island, may belong to this species as well. The behaviour of this form is here described as that of L. meeli. No specimens were collected for examination.

Both L. meeli and L. boulengeri differ from L. hecqui in that males of the latter species breed in harems and are usually able to fit inside a shell. L. hecqui does not excavate pits to conceal the shell(s) in the centre of the pit, as the other two do, and is often found in areas with a relative abundance of snail shells.

L. hecqui has a very large range and is found sympatrically with L. meeli (Poll, 1956). L. boulengeri seems to be its closest relative — this species was formerly synonymised with L. hecqui — and the two have a complementary distribution: L. boulengeri north of the Malagarasi and N. hecqui in the southern half of the lake, south of the Malagarasi and probably south of Kalemie. It is not yet known whether L. boulengeri occurs along the northwestern shore of the lake. Both L. hecqui and L. boulengeri, when breeding, are characterized by a pattern of dark irregular blotches on the body, while L. meeli, L. pleuromaculatus, and L. attenuatus normally have a single blotch on the centre of the flank. Stressed individuals of all five species exhibit a similar pattern of irregular blotches. L. hecqui differs from L. boulengeri by having a longer head and snout. L. boulengeri has a very distinct large blotch on the central part of the dorsal fin, but this marking is smaller or absent

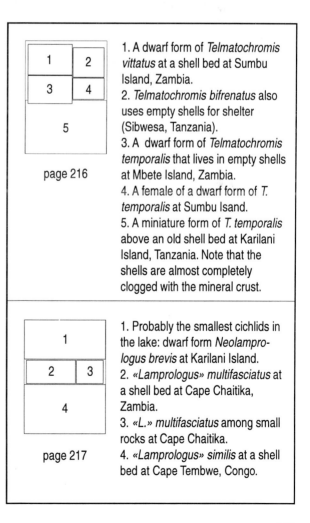

page 216

1. A dwarf form of Telmatochromis vittatus at a shell bed at Sumbu Island, Zambia.
2. Telmatochromis bifrenatus also uses empty shells for shelter (Sibwesa, Tanzania).
3. A dwarf form of Telmatochromis temporalis that lives in empty shells at Mbete Island, Zambia.
4. A female of a dwarf form of T. temporalis at Sumbu Isand.
5. A miniature form of T. temporalis above an old shell bed at Karilani Island, Tanzania. Note that the shells are almost completely clogged with the mineral crust.

page 217

1. Probably the smallest cichlids in the lake: dwarf form Neolamprologus brevis at Karilani Island.
2. «Lamprologus» multifasciatus at a shell bed at Cape Chaitika, Zambia.
3. «L.» multifasciatus among small rocks at Cape Chaitika.
4. «Lamprologus» similis at a shell bed at Cape Tembwe, Congo.

in *L. hecqui*. *L. meeli* differs from both species by a much narrower bony bridge between the eyes (interorbital).

The breeding biology of *L. hecqui* (often erroneously identified as *N. meeli*) has been studied only in the southern part of the lake (Sato & Gashagaza, 1997). It appears that a male can control up to four females in his territory. Each female removes sand from around a small group (maximum 10) of shells and defends this as her brooding territory. These territories are spaced 50 to 300 cm from each other and controlled by a single male. Some males seem too big to fit inside a shell; Poll (1956) records a maximum total length of 8.5 cm. Spawning takes place inside one of the female's shells and the male does not participate in the guarding of fry. Guarding females can easily be recognized by their very dark colour. The fry have a cryptic coloration and only three black spots in the dorsal are visible initially (see photos pages 212 and 213). When they grow older they acquire the typical blotchy coloration and are adorned with a very distinct ocellated spot in the dorsal fin. They are always found around empty snail shells.

The fact that *L. hecqui*'s breeding technique differs from that of *L. meeli* and *L. boulengeri* may be due to the abundant availability of empty shells within its range. In areas with only a few shells it may exhibit similar behaviour to that described for the other two species. A parallel can be seen in shell-brooders which are also found breeding in the rocky habitat and which change their normally monogamous breeding mechanism in rocky areas to polygynous in shell-covered habitats (Sato & Gashagaza, 1997).

Large shell beds

Single shells and areas with a low shell density are inhabited primarily by specialized shell-brooders; they seem to have little to offer to the facultative shell-brooders that use shells for protection of their offspring but are themselves too large to take refuge inside a shell. The reason may be that most of the single shells are already occupied by the better adapted species.

Large shell beds, however, are home to a number of cichlids, including both obligate and facultative shell-brooders, and such areas have a similar function to the thick mineral crusts on the rocks (see page 45) that give shelter to a host of small rock-dwelling cichlids. Not all shells in a shell bed are occupied and there seems to be an abundance of hideouts for juvenile cichlids as well.

There are several small species that are always found in large shell beds but none of them appears to be restricted to shells. The most common cichlid to be found in shell beds is *Telmatochromis brichardi*, a small cichlid which is also common in rocky habitats (see page 98). In the southern part of the lake, in Zambia, *T. brichardi*'s southern counterpart, *T. vittatus*, occupies the same shell-bed niche, while in the central part of the lake *T. bifrenatus* is also found on shell beds. All three species are also common, within their respective ranges, in the collections of shells guarded by «*L.*» *callipterus* males.

The maximum total length of *T. brichardi* and *T. bifrenatus* is about 6 cm, and such individuals can still find refuge inside shells when full-grown. *T. vittatus*, however, grows to a much larger size in the rocky habitat, up to 10 cm total length, but those found on shell beds never grow larger than about 6 cm.

We have seen previously that one shell-brooder, «*L.*» *speciosus*, has adapted its size to that of the available shells (page 207). In shell beds a number of normally rock-dwelling cichlids have adapted their size to be able to fit into shells for refuge. *T. vittatus* is one example, but another species of the same genus, *T. temporalis*, is very common on shell beds as well. The latter species has been exported as *T. burgeoni*, but current research by Tetsumi Takahashi (pers. comm.) suggests that there is hardly any difference between *T. temporalis* and *T. burgeoni*. *T. burgeoni* was described from specimens captured near Nyanza Lac in Burundi (Poll, 1942), and these may have derived from a population of the shell-dwelling form of *T. temporalis*, because the maximum total length is given as 7 cm — that of *T. temporalis* is 10 cm. The shell-brooding form is here regarded as a dwarf form of *T. temporalis*.

A shell bed at Karilani Island on the central eastern coast is inhabited by a dwarf form of the shell-dwelling form of *T. temporalis*, i.e. the *Telmatochromis* species found there is noticeably smaller than the ones found, for instance, in shell beds in Zambian waters. The reason for this miniaturization is, it seems, simply that the shells cannot accommodate larger fish. Another species, «*L.*» *brevis*, at the same bed appears to be much smaller than elsewhere. A male and female «*L.*» *brevis* usually occupy a single shell, but at Karilani the «*L.*» *brevis* males and females — of what must be the smallest cichlid species (or variant) in the lake — each have their own shell (see photo page 217). The shells are covered inside with such a thick crust of mineral deposits that only very small species can use them as shelters or breeding caves (see photo page 216).

Apart from the two miniaturized forms mentioned above, I also found *T. bifrenatus* at this bed, albeit in very small numbers.

The local configuration of the shells may have led to an extreme case of miniaturization at Karilani Island, a development which is seen at all shell beds but to a lesser degree. The uncommon situation at Karilani Island may be due to the absence of shell-brooders that cover or uncover shells. At other shell beds large cichlids sometimes dig holes in the bed and uncover shells that have no mineral deposits. Since the deposition of minerals proceeds very slowly and only on those objects that are in direct contact with the water, such uncovered shells provide living quarters for the smaller cichlids for a long period of time. Sato & Gashagaza (1997) found that *Neolamprologus cunningtoni* is one such digging species that sometimes digs its nest in shell beds. The uncovered clean shells are occupied by other species; in the case of the shell bed at Rumonge, Burundi, they found a dwarf form of «*L.*» *callipterus* (small males) taking refuge in shells near *N. cunningtoni* nests. Shells without calciferous deposits offer a larger interior and thus larger individuals can fit inside. In particular, species that are normally too large to fit inside a shell, e.g. *A. compressiceps* and *A. fasciatus*, are eager to get such clean shells.

The same researchers (Sato & Gashagaza, 1997) also found dwarf forms of *N. mondabu* at the Rumonge shell bed. *N. mondabu* normally breeds in the intermediate habitat where it excavates holes in the sand beneath rocks. In the shell bed they dig holes 6 to 7 cm deep and use the outer surfaces of shells on which to deposit their eggs. The wrigglers remain at the bottom of the pit or inside the shells in its wall, and are guarded by the female. Mobile fry hover above the pit and are guarded by both male and female. Males usually control two or three females.

These authors also found *Lepidiolamprologus pleuromaculatus* and *Telmatochromis dhonti* breeding in the shell bed at Rumonge. Both species are usually found in shallow and muddy habitats (see page 243), but at this site *T. dhonti* in particular exhibits interesting breeding behaviour. A territorial male guards a large shell of the bivalve *Iridina spekei* lying on top of several small snail shells. Up to three females may take up residence in the shells in the "burrow" formed beneath the bivalve shell. Spawning

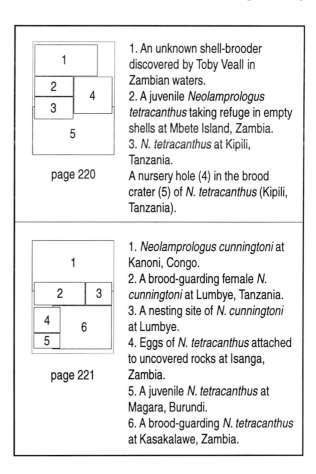

page 220

1. An unknown shell-brooder discovered by Toby Veall in Zambian waters.
2. A juvenile *Neolamprologus tetracanthus* taking refuge in empty shells at Mbete Island, Zambia.
3. *N. tetracanthus* at Kipili, Tanzania.
A nursery hole (4) in the brood crater (5) of *N. tetracanthus* (Kipili, Tanzania).

page 221

1. *Neolamprologus cunningtoni* at Kanoni, Congo.
2. A brood-guarding female *N. cunningtoni* at Lumbye, Tanzania.
3. A nesting site of *N. cunningtoni* at Lumbye.
4. Eggs of *N. tetracanthus* attached to uncovered rocks at Isanga, Zambia.
5. A juvenile *N. tetracanthus* at Magara, Burundi.
6. A brood-guarding *N. tetracanthus* at Kasakalawe, Zambia.

takes place inside the female's shell. The wrigglers are cared for by the female inside the shell, but soon after they become mobile disperse from the male's territory (Sato & Gashagaza, 1997).

Two closely related species, *«L.» multifasciatus* and *«L.» similis*, are often found associated with large shell beds. *«L.» multifasciatus* is restricted to Zambian waters whereas *«L.» similis* occurs in the remaining part of the southern half of the lake. *«L.» similis* is found as far north as Cape Tembwe along the western coast and Bulu Point on the east coast. Both species are also found among the rubble of rocky habitats but only at deeper levels (deeper than 25 metres of depth). Around shell beds both species can be found as shallow as 10 metres. They have a very interesting lifestyle as they are almost always found in colonies. *«L.» multifasciatus* and *«L.» similis* uncover buried shells at the edge of large shell beds and form colonies of up to 20 individuals. When the number of shells is becoming restrictive — each adult individual needs its own shell — a few split off and expose a new group of shells to create a new colony. This behaviour can be seen in the aquarium: these fishes dig the sand from around the cluster of shells provided and use it to form a rampart around their "caves".

«L.» multifasciatus and *«L.» similis* are among the smallest cichlids known to date: males grow no larger than 5 cm; females are rarely bigger than 3.5 cm. The individuals that live among the rocks have a larger maximum total length than those found in empty shells (5 as opposed to 4 cm). In shell beds they form communities of many pairs, with each individual having its own shell(s) to which to retreat when threatened. Spawning takes place inside the female's shell. After the fry have been free-swimming for about two weeks they leave the female's shell and wander through the cracks and holes of the shell bed and find refuge inside any shell they chose, even those occupied by an adult. They are tolerated by all members of the community until they reach maturity. Aquarium observations indicate that pairs normally remain together, but that new pair bonds may quickly be formed when the shells are rearranged. Even though *«L.» similis* has been described as a rock-brooding rather than a shell-brooding cichlid (Büscher, 1992c) it is very common at one shell bed near Cape Tembwe (Congo). On the other hand *«L.» multifasciatus* inhabits rocky biotopes as well (see photo page 217). Aquarium observations suggest that both species prefer shells to holes created by piling up pebbles and small stones.

«L.» similis has two bars on its head and can thereby be distinguished from *«L.» multifasciatus* which lacks bars on the head. Both species feed from the plankton drifting past their homes, but also eat any sizeable invertebrates found among the shells of their colony. The rock-dwelling individuals likewise feed on plankton as well as on invertebrates.

The sand

A sandy habitat is herein defined as a predominantly sandy bottom with less than one tenth of the area covered with rocks. Bare sandy areas support relatively few cichlids because the latter need some form of protection. Only those species living in large schools venture out over bare sandy bottoms. Since they forage in school formation they are infrequently encountered during dives over this type of habitat. The occurrence of fishes increases dramatically when a few rocks are present to provide shelter.

Sand-dwelling lamprologines

Neolamprologus tetracanthus and *N. cunningtoni* are sand-dwelling lamprologine cichlids found mostly in shallow water. Breeding takes place around rocks or shells, and breeding pairs are therefore often found at the edges of intermediate habitats. Both species are found throughout the lake, but geographical variation is small and known to be present only in *N. tetracanthus*.

N. tetracanthus can attain a maximum total length of 20 cm (males; females remain about 25% smaller). Populations in the northern part of the lake are characterized by a red margin to the dorsal fin, while the members of the southern populations have a yellow dorsal margin and also a yellow upper jaw. The juveniles of the northern variant have a black margin to the dorsal; this edging becomes red at maturity.

N. tetracanthus feeds on snails, insect larvae, and also fish (Poll, 1956). In the aquarium a single *N. tetracanthus* can keep a large snail population under control or completely eliminate it. Only tiny snails are eaten, and the shells are excreted after the gastric juices have digested their contents.

N. tetracanthus is not a monogamous substrate brooder as males tending two brood-guarding females have been found (Kuwamura, 1986a). The female selects a breeding site near rocks (or shells) and, typically, removes the sand from between and from beneath almost completely buried rocks. These rocks are often rather flat and, when the nest is ready, cover the breeding hole. If fewer than three adjacent rocks are uncovered, then a wall of sand, in a semicircular shape, can take the place of the missing rock(s). The eggs are deposited on the rocks and not infrequently are clearly visible. The wrigglers are kept in the bottom of the pit, under the rocks.

Sometimes the sand is coarse or mixed with small (bivalve) shells and is not much worked over by sand-grubbing species. Due to mineral deposition the sand in such areas is more or less caked together, and small holes dug in it do not cave in as they would in loose sand. *N. tetracanthus* sometimes makes use of such concrete-like sand, digging a small nursery hole in the wall surrounding the nest (see photo page 220).

The fry are guarded by both parents, in particular when the nest is open on one side. Guarding *N. tetracanthus* are readily recognizable as they have a pattern of broad dark bars on the body. The fry desert the nest at an early stage — large fry are rarely guarded by *N. tetracanthus*.

N. cunningtoni — this species was previously assigned to *Lepidiolamprologus* but Stiassny (1997) considers that it belongs to the large group of *Neolamprologus* — attains a much larger maximum total length than *N. tetracanthus*: almost 30 cm. *N. cunningtoni* is basically

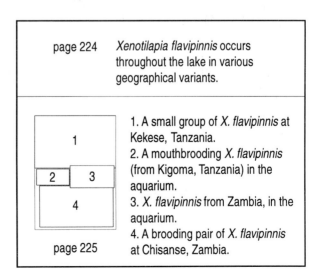

page 224	*Xenotilapia flavipinnis* occurs throughout the lake in various geographical variants.
1 / 2 3 / 4	1. A small group of *X. flavipinnis* at Kekese, Tanzania. 2. A mouthbrooding *X. flavipinnis* (from Kigoma, Tanzania) in the aquarium. 3. *X. flavipinnis* from Zambia, in the aquarium. 4. A brooding pair of *X. flavipinnis* at Chisanse, Zambia.
page 225	

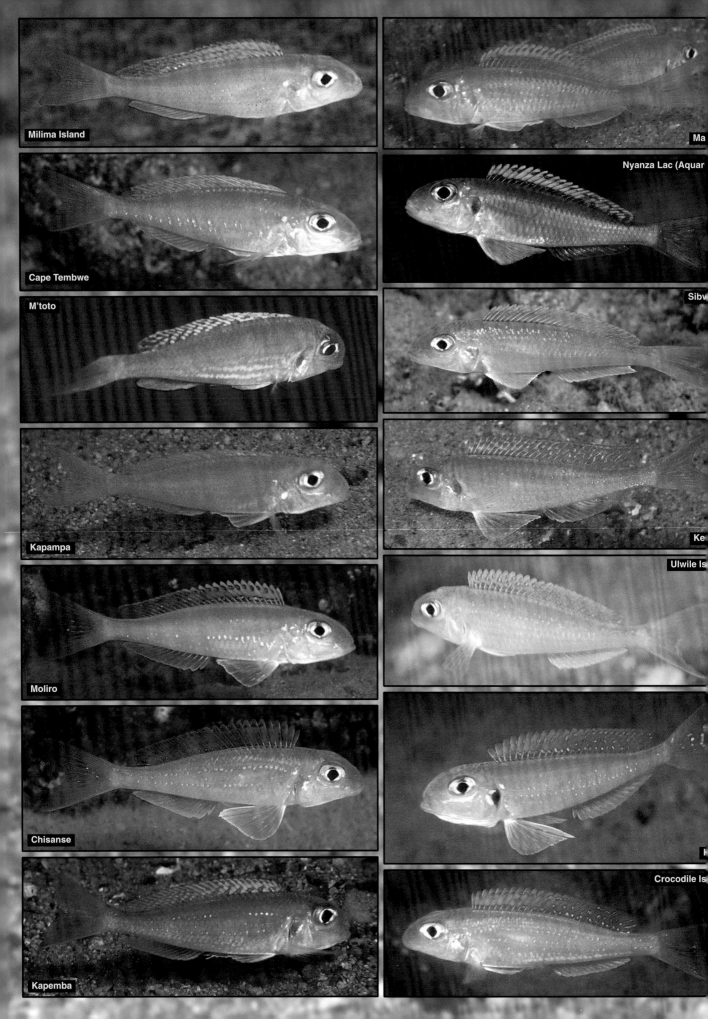

Milima Island

Ma

Cape Tembwe

Nyanza Lac (Aquar

M'toto

Sibw

Kapampa

Ke

Moliro

Ulwile Is

Chisanse

Kapemba

Crocodile Is

a piscivore, but feeds on almost anything it can get into its mouth. It can dive into the shells of shell-brooders and feed on the occupants; it has also been seen chasing sand-dwelling cichlids of the genus *Xenotilapia*.

Spawning takes place in a cave dug in the sand between rocks in a manner similar to that of *N. tetracanthus*. Since *N. cunningtoni* is larger than the latter, the nest is also larger and the rocks are farther apart. Spawns of *N. cunningtoni* are also larger: about 500 eggs per clutch as opposed to about 200 in *N. tetracanthus*. Breeding *N. cunningtoni* are very dark brown to almost black and readily distinguished from non-breeding individuals. Mobile fry hover above the nest and feed on plankton. Sometimes they are led through the habitat by the parents, several metres away from the nest.

Xenotilapia

The genus *Xenotilapia* is well represented in the sandy habitat. Among the species of this genus we find at least two different types of mouthbrooding. In one type, the maternal mouthbrooders, displaying males establish their territories close together in clusters; while in the other type, the biparental mouthbrooders, the fry are transferred from the female's mouth to that of the male halfway through the incubation period. The drawback of the latter system is that both male and female have to stay together, at least until the point of exchange. This behaviour has been mentioned earlier, e.g. as regards the goby cichlids (see page 27). In most cases male and female stay together for much longer than just a single breeding cycle.

Xenotilapia flavipinnis

X. flavipinnis is a small species — maximum total length about 11 cm — which lives in the shallow sandy habitat. Like all other members of *Xenotilapia* it feeds by pushing its mouth into the sand (or the muddy sediment covering it), scooping some of it up, and filtering it through the gills (large pieces of inedible material are spat out), and retaining anything edible. A unique characteristic of the members of this genus is the horizontal projection of the outer teeth on the lower jaw. These strangely-positioned teeth may facilitate penetration of the substrate (like the prongs of a fork) when the mouth is opened. The main items eaten are small crustaceans (in particular small *Cyclops*-like invertebrates that crawl on the bottom or hover just above it), tiny mussel shrimps, nematode worms, and insect larvae (Yanagisawa, 1986). *X. flavipinnis* always feeds from sandy substrates and its breeding territories are mostly situated there as well.

Sometimes pairs are unable to secure a breeding territory on the sand and in consequence some of them are found in the intermediate habitat. Yanagisawa (1986) found that such pairs consisted of smaller individuals than those found in nearby breeding areas on the sand. Non-breeding *X. flavipinnis* form large foraging schools and are often found in shallow water. When they become mature a male and female will secure a territory on the sand in the vicinity of the rocky habitat. Each pair needs a territory with a diameter of about three metres and all territories of the breeding group are contiguous, i.e. there is no free space in between territories. Feeding takes place inside the territory and pairs usually stay together within its boundaries for several consecutive broods .

Spawning takes place somewhere in the centre of the territory; a spawning pit is not excavated. The eggs — broods can number up to 40 fry — are first brooded by the female. After the larvae have hatched they are transferred to the male's mouth; this takes place 7 to 12 days after spawning. The female spits the larvae in front of the male who then quickly opens his mouth to let the larvae get in. The male continues to brood for another 5 to 6 days. Neither male nor female feeds during the mouthbrooding period. Yanagisawa (1986) found that the total brooding period takes about 15 days and that the fry are guarded for another two weeks before they leave the parents' territory. Parents that guard their offspring longer than others may have "foreign" juveniles, from neighbouring pairs, among their young.

X. flavipinnis has a lake-wide distribution and many geographical variants are known (see photos on page 224). The main difference in coloration — both male and female have the same colour — is in the markings on the flank, because almost all variants have yellow to orange dorsal fins. The variants found in the northern half of the lake all have a yellow dorsal fin and the one from Nyanza Lac (Burundi) is characterized by a yellow-orange colour on the snout and throat. At Cape Tembwe (Congo) *X. flavipinnis* has mother-of-pearl spots all over the flank and the variant at M'toto has a double yellow stripe on the anterior part of the flank. The form at Kapampa lacks yellow stripes, but further down the coast at Moliro (Congo) and Chimba (Zambia) *X. flavipinnis* has a single yellow horizontal bar above the belly. At Kapemba, however, the yellow stripe lies on the posterior part of the flank. This pattern is also found in the forms from the southeastern part of the lake. Along most of the Tanzanian shore *X. flavipinnis* has no markings other than a bluish body and a yellow dorsal. There is, however, a single population, between Kekese and Isonga, that is characterized by red dots in the dorsal fin. This variant was erroneously assigned to *X. boulengeri* in a previous publication (Konings, 1988) but now appears to be the most distinctive variant of *X. flavipinnis*. At Sibwesa, and further north along the Mahali Mountain range, the dorsal is orange, and some specimens have reddish dots in the dorsal as well.

Dream cichlids

One group of sand-dwelling species is characterized by gorgeously coloured males and silvery females. During the breeding season the males of at least four of these species acquire a beautiful and conspicuous breeding dress. Two of them, *Enantiopus* sp. "kilesa" and *Xenotilapia* sp. "ochrogenys ndole", have not yet been described; the other two are *Enantiopus melanogenys* and *Xenotilapia ochrogenys*. The latter two species have an almost lake-wide distribution, but neither one of them has been found in the range of *X.* sp. "kilesa", which extends along the western shore from Kalemie at least as far as the Kavala Islands. The "Kilesa" may have a wider range in the northwestern part of the lake.

X. ochrogenys and *X.* sp. "ochrogenys ndole" look very much alike and exhibit no obvious morphological differences although the latter appears to attain a slightly larger size. The coloration, however, differs, with male *X.* sp. "ochrogenys ndole" having three to five very conspicuous black dots on the flank. *X. ochrogenys* has not been found at Ndole, the locality where *X.* sp. "ochrogenys ndole" was first seen underwater. Poll (1956) records a southern form of *X. ochrogenys* which is characterized by black spots on the flank, clearly alluding to *X.* sp. "ochrogenys ndole". The range of *X.* sp. "ochrogenys ndole" encompasses most, if not all, of the Zambian shoreline. Although it may simply be a geographical variant of *X. ochrogenys*, it is here treated as a different species. This species/form has not yet been scientifically examined and

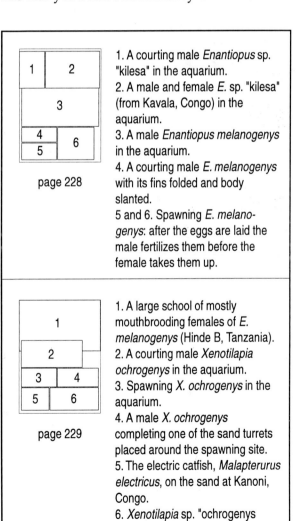

page 228

1. A courting male *Enantiopus* sp. "kilesa" in the aquarium.
2. A male and female *E.* sp. "kilesa" (from Kavala, Congo) in the aquarium.
3. A male *Enantiopus melanogenys* in the aquarium.
4. A courting male *E. melanogenys* with its fins folded and body slanted.
5 and 6. Spawning *E. melanogenys*: after the eggs are laid the male fertilizes them before the female takes them up.

page 229

1. A large school of mostly mouthbrooding females of *E. melanogenys* (Hinde B, Tanzania).
2. A courting male *Xenotilapia ochrogenys* in the aquarium.
3. Spawning *X. ochrogenys* in the aquarium.
4. A male *X. ochrogenys* completing one of the sand turrets placed around the spawning site.
5. The electric catfish, *Malapterurus electricus*, on the sand at Kanoni, Congo.
6. *Xenotilapia* sp. "ochrogenys ndole" (from Ndole Bay, Zambia) in the aquarium.

occurs in an area where there are several "new" species, because it is a relatively new part of the lake (see also page 14).

E. melanogenys and *E.* sp. "kilesa" are very slender, elongate, cichlids with a long snout. *E.* sp. "kilesa" differs from *E. melanogenys* in the shorter lower jaw, the shorter snout, and the yellow colour on the chin. All known species of the genus *Xenotilapia* likewise have slender bodies, but they differ in having rounded heads with a very steep upper snout profile. The two species of *Enantiopus* reach a maximum total length of about 16 cm; the maximum size of *X. ochrogenys* is about 12 cm, although aquarium specimens of *X.* sp. "ochrogenys ndole" can attain a maximum total length of 14 cm. There are no distinct geographical races known for these four species.

These sand-dwelling cichlids forage in large schools over the sandy substrate. The maximum depth recorded for *E. melanogenys* is 40 metres, but during the breeding period they are found in much shallower water. *X. ochrogenys* is found in shallow water throughout the seasons. When not breeding, males have the same sandy, silvery, coloration as females, which gives them optimum camouflage on the sandy bottom. Most individuals of a single school remain together in that school throughout their lives. A school is probably formed at the moment fry are released simultaneously by mouthbrooding females. The youngsters grow and breed together until they die after about three years.

These species are all mouthbrooders in which only the female incubates the eggs and larvae. Breeding probably takes place throughout the year, but the highest activity occurs during the rainy season, from December to May. During the rest of the year most of the schools move around and forage in different areas. *X. ochrogenys* normally occurs in small schools of up to 30, but schools of *Enantiopus* can number in the hundreds. Because there are so many fish in one school together they do not have to seek shelter when they want to breed. Since most members of the school are the same age they all attain their reproductive phase at about the same time. It seems that *X. ochrogenys* has attached itself to the breeding colonies of the *Enantiopus* since they are usually found breeding together (Brichard, 1978).

During the breeding season males defend territories and try to persuade females to spawn in their nests. The actual breeding may take place in bouts possibly lasting less than a week, and during such periods the fish do not eat but concentrate on spawning. The breeding season begins when the males start staking out territories in the sand. A male *E. melanogenys* digs a flat, saucer-shaped, territory with a diameter of about 50 cm. In the centre of the territory he digs a small pit with a diameter of about 15 cm. This will be the nest in which spawning will take place. The territory of *X. ochrogenys* and *X.* sp. "ochrogenys ndole" is rather peculiar; it consists of three to eight turrets built by heaping sand. These are erected in a circle around the spawning site, which is a round, saucer-shaped, pit with a diameter of approximately 10 cm.

The territory of *E.* sp. "kilesa" is a very interesting "mixture" of the former two types. In the aquarium a male will occupy a large area in which he makes several nests! These nests are shallow pits with a diameter of about 15 cm. Around these nests the male heaps sand turrets — sometimes more than 20 — all over his territory. Spawning usually takes place in a single nest, but the next time around the couple may spawn in another. So *E.* sp. "kilesa" not only looks like a hybrid between *X. ochrogenys* and *E. melanogenys* in terms of morphology and colour, but in addition its nest-building behaviour seems to be a mixture of that performed by the other two species. If there exists anywhere among cichlids a species that could have originated by the hybridization of two other species, then *E.* sp. "kilesa" is a prime candidate for the role.

The nests of these sand-dwelling cichlids are meticulously cleared of small pebbles which are larger than the prevailing sand grains. The reason why males go to the trouble of removing larger grains is obvious: so that the female cannot mistake a tiny pebble for an egg when she collects them after spawning. On an evenly-structured sand nest the eggs are more conspicuous and can be collected faster.

When the territory is ready the male starts courting. Here we notice some differences between *Enantiopus* and *Xenotilapia*. The male *X. ochrogenys* attracts a female to the nest with his fins fully erected, except the first part of the dorsal fin which is only half erected. The male *Enantiopus*, on the other hand, attracts females in a most peculiar fashion. Rivals are chased with all fins erect, but females are seduced to the nest with all fins clamped! The only thing extended is the buccal cavity, as in *Xenotilapia*. Furthermore the male lies almost completely on his side when attracting the other sex. In many other species the female is directed to the correct spawning place by the male holding his egg-dummies in the centre of the nest; but in *E. melanogenys* the female may equally be enticed to egg-laying by the male imitating a mouth full of eggs!

Males are not aggressive among themselves and the female is not chased when she ignores the courtship of a male. As a female passes over the different territories, which abut one other, the males, one by one, lie on their sides and try to seduce her. The males, however, remain in their territories.

When a female is ready to spawn she responds to the display of the male. As soon as she enters the nest, the male circles around her with fins erect — in *Enantiopus* as well as in *Xenotilapia*. Males are so excited that they first chase all other fishes away from the nest. Meanwhile, the female remains motionless in the nest. After clearing the site of intruders, the male enters the nest with fins erect and buccal cavity extended. He gently nudges the female in the hind part of her body and so encourages her to start circling around. During circling the male vibrates his extended buccal cavity. After two to three uninterrupted rounds the female suddenly slows down and lays some eggs. The male too stops circling and waits impatiently for the female to move from the site and leave the eggs to be fertilized. One to eight two-millimetre-long eggs are deposited at a time. As soon as the female moves forward the male shoots over the eggs and releases milt over them; the eggs are thus fertilized outside the female's mouth. At the end of the spawning cycle the female ceases releasing any more eggs, but the ritual goes on long after that. Every time she moves forward the male shoots like an arrow over the barren sand. Sometimes a male can be so excited that he bolts over the female when she does not move fast enough.

Clutch size varies between 30 and 80 eggs for the two species of *Enantiopus*, and between 10 and 40 for the other two. A male may spawn with several females on a single day, and a female may have her eggs fertilized by two or three different males. Within a short period of time all the females are mouthbrooding. Large schools of brooding females are often found near the breeding arenas of the males, and they remain in the shallow water until, after about three weeks, the rather small fry are released simultaneously. The latter are mature after about a year and can spawn during the next breeding season.

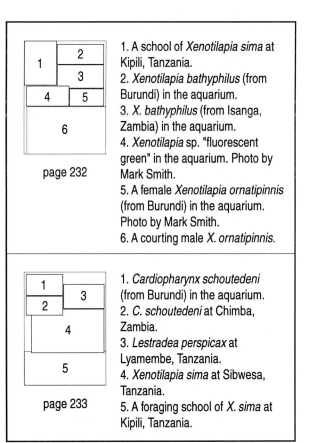

1. A school of *Xenotilapia sima* at Kipili, Tanzania.
2. *Xenotilapia bathyphilus* (from Burundi) in the aquarium.
3. *X. bathyphilus* (from Isanga, Zambia) in the aquarium.
4. *Xenotilapia* sp. "fluorescent green" in the aquarium. Photo by Mark Smith.
5. A female *Xenotilapia ornatipinnis* (from Burundi) in the aquarium. Photo by Mark Smith.
6. A courting male *X. ornatipinnis*.

page 232

1. *Cardiopharynx schoutedeni* (from Burundi) in the aquarium.
2. *C. schoutedeni* at Chimba, Zambia.
3. *Lestradea perspicax* at Lyamembe, Tanzania.
4. *Xenotilapia sima* at Sibwesa, Tanzania.
5. A foraging school of *X. sima* at Kipili, Tanzania.

page 233

Xenotilapia bathyphilus

X. bathyphilus was regarded as a subspecies of *X. ochrogenys* before Poll revised this genus in 1986. Both species are, however, sympatric. As can be anticipated from its name, *X. bathyphilus* lives in the deeper parts of the lake, and depths ranging between 20 and 100 metres have been recorded (Poll, 1956). It bears a close resemblance to *X. ochrogenys* and has a lake-wide distribution. *X. bathyphilus*-like cichlids have been maintained in aquarists' tanks, with some actually being labelled *X. bathyphilus*, while others have received trade names such as "Xenotilapia Isanga". Males of this form, collected near Isanga (Zambia), have a steel-blue cast on the upper part of the flank and a yellow dorsal fin.

X. sima or *X. boulengeri*

Xenotilapia sima and *X. boulengeri* are very closely related and there are no distinct characters which would unequivocally set these two species apart. Both are found in shallow water, and both have a lake-wide distribution. Recently Takahashi & Nakaya (1997) revised *X. sima* and *X. boulengeri* and concluded that there are two different species, both with a lake-wide distribution. However, the characters given to distinguish these two putative forms are all overlapping and I remain unconvinced that *X. boulengeri* is a separate species to *X. sima*. It seems remarkable that one of the two is seen during almost every dive, while the other has apparently never been seen alive. And until such time as behavioural data become available that clearly demonstrate that these are indeed two different species, I regard *X. boulengeri* as a junior synonym of *X. sima*.

X. sima attains a maximum total length of about 16 cm and is quite common in the shallow sandy habitat. It is almost always seen in foraging groups numbering between five and more than a hundred individuals. It feeds by filtering the upper layer of the sandy substrate and its diet consists of insect larvae, small snails, and tiny clams (Poll, 1956). It breeds on sandy patches in the intermediate habitat but sand-scrape nests or other forms of spawning sites are not constructed. Mouthbrooding females stay together and probably release their offspring simultaneously.

Kuwamura (1986a; 1987b) and others have described *X. boulengeri* as a biparental mouthbrooder, but it is virtually certain that they confused this species (i.e. *X. sima*) with *X. flavipinnis*.

Xenotilapia ornatipinnis

The most conspicuous features of *X. ornatipinnis* (maximum total length about 13 cm) are the enormous eyes and the silvery mid-lateral stripe on the male's flank. The eye is not perfectly round but oval; this allows "more" eye to fit into the available space and the construction of the skull. Large eyes have a better light perception than small ones, and such enormous eyes can be advantageous under subdued light conditions such as exist in deep or murky habitats. In addition to the large eyes, which improve the vision of the individual, the silvery stripe on the male's flank probably serves as a recognition mark to enable females to find their mates in deep or turbid water. Poll (1956) reports that most specimens were collected in deep water, some from as deep as 110-160 metres, but also records collection of this species at a depth of five metres. Its diet consists of insect larvae and small mussel shrimps. Poll (1956) also found sand grains in stomach contents.

X. ornatipinnis, which has a lake-wide distribution, is a maternal mouthbrooder; females lack the silvery stripe and have a row of five to seven ill-defined black dots in the dorsal fin. In the aquarium males do not construct a spawning site and females are courted on the sand as well as on rocks.

Xenos with a nose

Recently a new species of *Xenotilapia*, *X. nasus*, has been described from the northern section of the lake (De Vos *et al.*, 1996). *X. nasus* is characterized by an underslung mouth and a snout which appears nose-like, similar to that seen in *Ophthalmotilapia nasuta*. No other described species of the genus *Xenotilapia* has such a snout, and very few others have only 7-8 soft rays in the anal fin, another character of *X. nasus.*

A few years ago a sand-dwelling cichlid was caught in Zambian waters and exported as "Xenotilapia Fluorescent Green". This form too has a nose-like protruding snout and 7-8 soft rays in the anal fin. In contrast to *X. nasus,* the tail of the Fluorescent Green is not strongly forked and does not have pointed lobes. It seems, however, that these two forms are

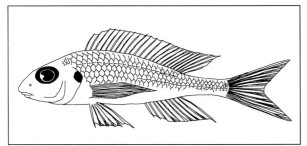

Xenotilapia nasus (after De Vos, Risch & Thys v. d. Audenaerde, 1996).

closely related, but very little is known of the natural behaviour of *X.* sp. "fluorescent green".

X. nasus has been found only in the extreme northern part of the lake, in Burundi as well as in the Congo. Most individuals have been caught at levels between 30 and 68 metres. Stomach contents inventories suggest that it feeds on detritus as well as on plankton.

Sand-dwelling Ectodini

This group of sand-dwelling cichlids consists of several silvery-coloured species that are assigned to three different genera: *Cardiopharynx schoutedeni, Ectodus descampsi, Ectodus* sp. "descampsi ndole", and *Lestradea perspicax.* the latter species formerly consisted of two subspecies, *L. p. perspicax* and *L. p. stappersi,* but the latter has subsequently been raised to full specific status (Poll, 1986). The two forms are nowhere sympatric, which would prove that there are indeed two species involved. The sole difference between them is the shape of the teeth (conical or bicuspid), and this is not accepted here as an adequate indication of two different species. Tooth shape is often dependent on

available food source(s) and other environmental factors, and it is highly unlikely that it plays a similar role in mate recognition to male coloration. *Lestradea stappersi* is therefore here regarded as a junior synonym of *L. perspicax.*

The cichlid now called *Ectodus* sp. "descampsi ndole" has previously been assigned to *E. descampsi* (Konings, 1988). The holotype of the latter species was caught near Moliro (Congo), not far from Ndole Bay (Zambia); nevertheless the morphology of the undescribed form, which has been recorded only from Ndole Bay but may possibly have a wider distribution along the Zambian shore, is sufficiently different from that of the holotype (and other specimens, collected in Burundi) that it is here treated as a distinct species. It has a deeper body (3.5 times in

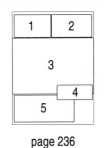

page 236

1. *Ectodus* sp. "descampsi ndole" in Ndole Bay, Zambia. Photo by Horst Dieckhoff.
2. *Ectodus descampsi* (from Burundi) in the aquarium.
3. A courting male *Lestradea perspicax* in the aquarium. Photo by Mark Smith.
4. A male *Grammatotria lemairii* in breeding colours.
5. *G. lemairii* at Katete, Zambia.

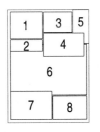

page 237

1. A plaque of eggs of *Boulengerochromis microlepis* on a rock at Cape Kachese, Zambia.
2. Eggs of *B. microlepis* deposited on sand at Kipili, Tanzania. Wrigglers of *B. microlepis* are kept in various small pits, sometimes in the rocky habitat (3. Kekese, Tanzania), or sometimes on the sand (4. Kipili).
5. Free-swimming *B. microlepis* fry are very vulnerable to predation.
6. *B. microlepis* not only tastes good but is also a gorgeous cichlid.
7. Most of the *B. microlepis* fry fall prey to other fishes during their first few days of free-swimming.
8. *B. microlepis* juveniles are guarded for as long as their parents live. In this case the "fry" are almost a year old but are still being guarded.

standard length vs. 4 to 5 times in *E. descampsi*) and the spines of the dorsal fin are noticeably longer than the soft rays, giving the fin a flag-like appearance.

It should be mentioned here that there are at least two other species that are not limited to rocky substrates which have developed a new form in the same area: *Xenotilapia* sp. "ochrogenys ndole" and *Microdontochromis rotundiventralis*. And, as with *E. descampsi*, the "mother-species" of the other two (*X. ochrogenys* and *M. tenuidentatus*) are found at Moliro and further north. This is a very interesting phenomenon that needs further investigation as regards the mechanics of speciation in this part of the lake.

The four species mentioned above all have a perfect camouflage: a silvery body that reflects the sandy bottom, making it difficult for predators to make them out. They are usually found in large schools, foraging from the bottom or at some distance from it, feeding on plankton. All four species are excellent sand divers: in order to escape an imminent predator attack they thrust themselves into the sandy substrate. Another common character is the black spot in the dorsal fin which is already visible in fry and remains in the adult fish, except in *Lestradea* where the spot disappears a few days after the fry have been released (Eysel, 1990).

Cardiopharynx schoutedeni and *Cyathopharynx furcifer* have a few remarkable anatomical features in common. Like *C. furcifer*, *C. schoutedeni* has a round or heart-shaped lower pharyngeal bone densely set with small teeth. This feature caused Greenwood (1983) to synonymise the genus *Cardiopharynx* with *Cyathopharynx*, but Poll (1986) pointed out that pharyngeals of this type are also found in cichlids from outside Lake Tanganyika, and thus cannot be seen as a unique character uniting into a single genus all species possessing it .

Examination of stomach contents has revealed a spectrum of micro-organisms and unicellular algae together with a considerable amount of very fine sand — not unlike the stomach contents of *C. furcifer*. *C. schoutedeni* is found on pure sand bottoms and occurs throughout the lake. On every occasion I was able to observe this species most of the individuals of the school

swam in mid-water about 50 to 100 cm above the sand. They were feeding on plankton, probably phytoplankton. The sand grains found in their very long intestines (almost four times as long as the fish's total length) are probably ingested when they scoop diatoms from the bottom at times when there is no plankton. Males can attain a maximum total length of 15 cm; females remain a few centimetres smaller.

Sexually active males have black pelvic fins and a black throat. They dig a small crater with a diameter of about 15 cm in the sand and try to persuade females to mate with them. I have not seen breeding arenas; male's nests are spaced at least three metres apart. Spawning probably takes place very early in the morning.

Ectodus descampsi is a relatively small species growing no larger than about 10 cm, and is more of a sand-dweller than *C. schoutedeni*. It feeds on anything it can find in the sandy substrate, which is filtered and chewed. It has a short intestine (about as long as the fish's total length) and the main fare is insect larvae.

Territorial males dig craters with a diameter of about 20 cm in the sand. Spawning probably takes place during the twilight hours, as has been observed in specimens kept in the aquarium. The large spot in the male's dorsal is surrounded by a light-blue halo which makes it very conspicuous, including under subdued light conditions. Clutch size varies between 15 and 35 eggs, and these are brooded for about three weeks. The tiny fry already have the characteristic spot in the dorsal the moment they are released.

Lestradea perspicax feeds on very small food particles, and is, like *C. schoutedeni*, a diatom-eater. Both species are often found together, in particular when feeding from the plankton above the bottom. *L. perspicax* also feeds from the silty layer covering the sandy substrate, and sand appears to be a common ingredient of its intestinal contents (Poll, 1956). Males construct large sand-scrape nests in the sand and are a bluish colour with a beautifully marked dorsal fin. Maximum total length is about 14 cm.

Grammatotria lemairii

Grammatotria lemairii is the largest sand-grubbing cichlid in the lake and can attain a total length of about 18 cm. It is common over sandy substrates all over the lake, and can easily be recognized by the black spot on its caudal peduncle. *G. lemairii* feeds on invertebrates (insect larvae, small snails and clams, crustaceans, etc.) which it finds in the sand. While most other sand-filtering species (*Xenotilapia* and *Callochromis*) sort through the upper layer of the substrate, *G. lemairii* digs deeper and searches for food a few centimetres below the surface. In order to reach this level they first blow the upper layer away, in a manner similar to that seen in *Xenotilapia*, and thrust their heads into the sand. The head, which sports a very long snout, can be completely buried while sand is collected.

Observations in captivity indicate that males are not territorial and do not construct spawning sites. *G. lemairii* is sand-coloured when not breeding but courting males have a black patch on the head and nape. The breeding male also exhibits an indistinct metallic hue and some intensified markings on the body. Displaying males sequester ripe females from the school and mate on the spot. Mouthbrooding females join the school and incubate the eggs for almost four weeks before they release the fry.

The world's largest cichlid

Boulengerochromis microlepis is the largest cichlid known to mankind and can weigh more than three kilos. Some specimens are longer than 70 cm, and the "Kue" or "Nkuhe", as it is known locally, is valued for its firm meat and good taste. *B. microlepis* occurs throughout the lake and in the Zambian section of the lake it is one of the species sport anglers vie to catch. They call it "Yellow Belly", referring to the yellow colour of male and female.

B. microlepis is a piscivore when mature; juveniles of up to 20 cm may also feed on larger invertebrates. It seems to be most common in the deeper sandy habitats and adult individuals are seen in the shallow water only when they want

to breed. In the deeper parts of the lake they pursue the herring-like fishes of the open water, and other abundantly available fishes such as schooling *Xenotilapia* species, on the sand.

Spawning takes place in the shallow intermediate habitat and it is here that most sightings occur. Very frequently a pair can be seen together, either preparing to spawn or guarding their offspring. The female is about a quarter to a third smaller than the male, who has, on average, a total length of about 60 cm.

B. microlepis is an open-substrate brooder in which both male and female guard the fry. A few days before the pair spawn the terrain is examined and a suitable location selected. Kuwamura (1986b) found that certain spawning sites in an area seem to attract *B. microlepis* more than others, as he observed spawnings by up to four different pairs at exactly the same site. *B. microlepis* does not always use flat rocks on which to deposit eggs — I have seen eggs on shells as well as on open sand. All nests are, however, close to rocks. If the spawning is to take place on sand a shallow crater is dug by the female before the actual mating. The female

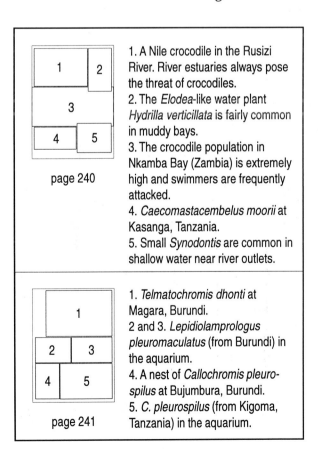

page 240

1. A Nile crocodile in the Rusizi River. River estuaries always pose the threat of crocodiles.
2. The *Elodea*-like water plant *Hydrilla verticillata* is fairly common in muddy bays.
3. The crocodile population in Nkamba Bay (Zambia) is extremely high and swimmers are frequently attacked.
4. *Caecomastacembelus moorii* at Kasanga, Tanzania.
5. Small *Synodontis* are common in shallow water near river outlets.

page 241

1. *Telmatochromis dhonti* at Magara, Burundi.
2 and 3. *Lepidiolamprologus pleuromaculatus* (from Burundi) in the aquarium.
4. A nest of *Callochromis pleurospilus* at Bujumbura, Burundi.
5. *C. pleurospilus* (from Kigoma, Tanzania) in the aquarium.

initiates spawning by starting to lay rows upon rows of eggs which are fertilized by the male when the female leaves the nest to lay rows in a different direction. Although all intruders are chased from the immediate environs of the nest, almost all aggression is directed towards conspecifics which are chased up to 10 metres away from it. Kuwamura (1986b) notes that spawning can take almost two hours and that between 5,000 and 12,000 eggs are deposited.

It takes about three days before the eggs hatch and the larvae are then immediately transferred by the female to a small pit nearby. The male does not take part in this initial transport, but both parents guard their wrigglers. During the next five days, before the larvae become free-swimming, they are frequently transferred between pits, sometimes as far as 10 metres away. This activity is performed by both parents and its function is probably to outsmart predators that find prey by smell, in particular catfishes that hunt at night. When the smell of palatable young fish is spread out over an area of 100 square metres it becomes a lot more difficult to locate the larvae which are all huddled together in a pit with a diameter of about 15 cm.

Eight days after spawning the fry become mobile and hover over the nest feeding on plankton. This is the most vulnerable time as the parents cannot control the movements of more than 10,000 fry at the same time. The fry stay together in a dense cloud guarded by a parent on each side. However, predators make quick incursions from all sides into the multitude of fry and decimate them to a considerable extent. Most of these early-stage predators are rock-dwelling cichlids and the pair therefore lead their fry to deeper and more sandy habitats.

It has been found in captivity that the female ceases eating about a week before spawning, nor does she eat during the period that she has fry (Fohrman, 1994). Poll (1956) and Kuwamura (1986b) both examined breeding females and found that they all had empty stomachs and sometimes even males have empty stomachs. Poll also noticed that the viscera of large specimens that were sexually ripe were almost completely absorbed and thus non-functional. He suggested that *B. microlepis* probably breeds once in a lifetime. Although we know that the viscera in other fish can completely regenerate after having been in an inactive state, I concur with Poll that *B. microlepis* is a semelparous species, i.e. all of its offspring are produced at a single spawning.

Fry-guarding parents are far too occupied with fending off predators to be able to hunt themselves. One could argue that after some time the fry are abandoned and the parents start to eat again, but on several occasions I have seen *B. microlepis* defending offspring which had an average length of about 15 cm (and thus an age of at least 9 months!). It was quite obvious that these parents had not been able to spawn again, and since they were very emaciated I doubted if they ever would.

The shallow sediment-rich habitat

The shallow sediment-rich habitat is herein characterized as a mixture of sand/mud and rocks on a gradually shelving shore. The rocks are in very shallow water — the shallow sediment-rich habitat is no deeper than about three metres — and covered by a layer of muddy sediment. The mud derives from a nearby river, and such habitats are commonly found near river mouths in shallow bays. In addition water plants are found in such habitats, the four most common species of higher plants being *Vallisneria spiralis, Ceratophyllum demersum, Myriophyllum spicatum*, and *Hydrilla verticillata* (determination according to Kasselmann, 1998). Although algae-feeding cichlids are sometimes found in such sediment-rich habitats they are much more common in other rocky habitats, and for that reason have been discussed earlier.

Substrate brooders

The shallow water and the heavy accumulation of sediment prevent many cave brooders from excavating tunnel nests under rocks, and only two species are mentioned here: *Telmatochromis dhonti* and *Lepidiolamprologus pleuromaculatus*. *T. dhonti* is better known as *T. caninus*, but the latter name is now regarded as a junior synonym (Poll, 1986). *T. dhonti* has a very large range and has been found mostly in the northern section of the lake, although it has been recorded from Moliro (Congo) and Utinta (Tanzania) as well. It can be found in many different habitats but is very common in very shallow water. It is often found in turbid water, and on one occasion has even been found in the Lukuga River (Poll, 1956). *T. dhonti* is a predator, feeding on anything it can swallow. In the aquarium it is a rather aggressive species, attacking other species as well as conspecifics.

Males, which attain a maximum total length of about 12.5 cm, defend a breeding territory in which several females, which are about half the male's length, have their nests. A dwarf morph of *T. dhonti* has been found on shell beds using empty shells to breed (see page 219).

Lepidiolamprologus pleuromaculatus also breeds in shell beds, but, like *T. dhonti*, is more often found in the shallow sediment-rich rocky habitats. *L. pleuromaculatus* has a restricted range and has been found only in the northernmost section of the lake. Almost all specimens exported for the hobby derive from the area near Bujumbura (Burundi).

The maximum total length of *L. pleuromaculatus* is approximately 11 cm; nevertheless it is a piscivore feeding on fry of other species — sometimes those of shell-brooding cichlids. Males occasionally have more than one female each (Sato & Gashagaza, 1997) and defend a small group of rocks as their territory. The eggs are not well-hidden and are deposited on a rock (or shell). After they hatch the wrigglers are transferred to a sand pit where they are cared for by the female. Clutches average about 50 eggs.

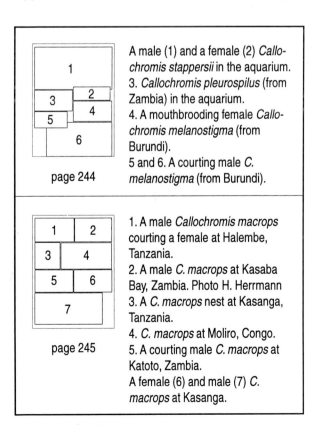

A male (1) and a female (2) *Callochromis stappersii* in the aquarium.
3. *Callochromis pleurospilus* (from Zambia) in the aquarium.
4. A mouthbrooding female *Callochromis melanostigma* (from Burundi).
5 and 6. A courting male *C. melanostigma* (from Burundi).

page 244

1. A male *Callochromis macrops* courting a female at Halembe, Tanzania.
2. A male *C. macrops* at Kasaba Bay, Zambia. Photo H. Herrmann
3. A *C. macrops* nest at Kasanga, Tanzania.
4. *C. macrops* at Moliro, Congo.
5. A courting male *C. macrops* at Katoto, Zambia.
A female (6) and male (7) *C. macrops* at Kasanga.

page 245

Sand cone builders

The genus *Callochromis* consists of four species of sand-grubbing cichlids found in the shallow waters of the lake. *C. pleurospilus* and *C. stappersii* are the smallest and can grow to a size of a little over 11 cm. *C. macrops* and *C. melanostigma* are bigger and can attain a maximum total length of about 16 cm. *Callochromis* form large foraging schools, as do all other sand-dwelling species of a small size.

Callochromis species are generalized carnivores, feeding on anything they find in the sand. The upper layer of the sand is randomly chewed and filtered for food. Their diet consists of insect larvae, small snails, tiny mussel shrimps, worms, and other invertebrates. In the aquarium they leave the bottom strewn with small pits, very similar to those seen in the lake.

Callochromis species are lek-breeders, which construct their sand-turret nests near rocks or water plants. Most breeding arenas are found in water less than two metres deep. Male *C. pleurospilus* (and possibly *C. stappersii*) build small sand-castle nests which have a diameter of about 15 cm and an elevated rim. The larger *Callochromis* species build larger and higher nests which usually consist of a heap of sand about 20-30 cm above the bottom and have a flat spawning platform on top. Sometimes the nest is built on top of a small rock, completely encircling it, and with approximately the same overall height as regular nests. The nests are normally about two to three metres apart. The fact that the nests are "built-up" makes them more visible to females, which stay very close to the bottom. The ripe females huddle together under the cover of water plants or at the edge of a more rocky section of the habitat. Mouth-brooding females are often found solitary or in small groups, and may also hide in the vegetation of the habitat.

Males court females with great zeal and try to lead them to their spawning sites. Other males are vigorously expelled from the vicinity of the nest. The male shows the female the place where she has to deposit eggs by folding his anal fin in such a manner that the large, orange-red, spot on this fin resembles an egg in three dimensions (!), and holding it in the centre of the spawning platform. A ripe female responds to the courtship display of the male by entering his territory. The female deposits some eggs on the platform, then turns around and picks them up. At the same time the male holds his folded anal fin in front of the female's mouth and discharges his milt. This is taken in by the female together with the eggs, which are thus fertilised. The clutch size of the smaller *Callochromis* ranges between 15 and 50 eggs and that of the two larger species between 25 and 60 eggs.

C. pleurospilus has a lake-wide distribution but the other three species have a restricted range. The types of *C. stappersii* were caught at Tulo and Kilewa Bay (Tanzania) and the aquarium population derives from an unknown locality in the Congo. *C. pleurospilus* is known to occur in geographical variants, and three different forms have been exported for the hobby: the red-dotted variant from Burundi, a red variant from Zambia, and the so-called "Callochromis Greshakei" or "Callochromis Rainbow", which has a wide red margin to the dorsal and a bluish body, and comes from Kigoma (Tanzania).

C. macrops has a very wide range and is found almost anywhere around the lake except in the extreme north, north of Nyanza Lac (Burundi). It is found sympatrically with *C. pleurospilus*, but the latter seems to be restricted to extremely shallow water, shallower than one metre. *C. macrops* is not found together with *C. melanostigma* and this species may just be a geographical variant of *C. macrops*. *C. melanostigma* is found only along the northern shores; north of Nyanza Lac and Cape Caramba (Congo).

Several geographical variants of *C. macrops* are known. The form at Kigoma resembles *C. melanostigma* but lacks the black markings in the dorsal fin. Many distinct variants occur along the Zambian shoreline. It is remarkable that these variants have such a variable coloration in relation to the nearest different form. The so-called "Red Macrops" from Zambia is found in Nkamba Bay and has a golden-red hue all over its body. An even darker red is found on the sides of the Ndole and Sumbu variants. The Ndole variant has black markings on a blue dorsal fin whereas the Sumbu variety bears additional markings on the cheeks. Not far from these red forms a purple-coloured form occurs in Kasaba Bay. It is remarkable that such geographical variation exists in a sand-dwelling species, and this may suggest that the breeding habitat — shallow rocky shores — plays an important role in the segregation of the different populations.

Ctenochromis horei

C. horei is probably the most common species in the shallow sediment-rich habitat and occurs throughout the lake. Males can attain a maximum total length of about 18.5 cm and females about 14 cm. It is an omnivore, feeding on water plants as well as on invertebrates and small fishes. C. horei is most often found in vegetated areas and males often clear areas in plant beds (the most common plants in such beds are hornwort and Vallisneria) as their spawning sites. These sites have a diameter of about 30 cm and are very conspicuous when one snorkels over the plant beds. Females are commonly found in small foraging groups among the plants or on the sand.

Breeding in C. horei takes place during all seasons of the year, and the population density at most localities is too high to allow every adult male to possess a territory at the same time. There appears to be continuous conflict among the males to obtain access to prime breeding sites, and at any given time only a few large males exhibit breeding coloration, indicating that they have succeeded in their objective. Mouthbrooding females are solitary and hide among the plants, staying very close to the bottom. Sometimes they hide between rocks.

The clutch size of C. horei ranges between 25 and 100 eggs, but Sato (1986) found that at Uvira (Congo) 15% of the eggs brooded by C. horei females were catfish eggs. Synodontis multipunctatus is a brood parasite that allows its eggs to be brooded by cichlids. A male and female catfish are stimulated to spawn by the smell of spawning cichlids. When the cichlid female has deposited some eggs and is turning around to pick them up, the catfishes intervene by quickly gobbling up the eggs just laid and replacing them with a batch of their own. The cichlid female is tricked into picking up the smaller catfish eggs, believing that they are her own, and continues spawning. After spawning she broods both types of eggs, her own and those of the Synodontis. The catfish eggs, however, grow much faster than the cichlid's, and once the baby catfishes have absorbed their yolk sacs they start feeding on their host's eggs and larvae (see photo page 248). For as long as the female continues to brood, the catfishes remain inside her mouth, and if this period outlasts the cichlid food supply then the larger hungry catfishes start eating their smaller siblings. If the female completes her regular brooding period then there are usually only a few catfishes left.

It has been found that in the aquarium S. multipunctatus will accept other mouthbrooding cichlids as host, and they have been bred with Victorian as well as Malawian mouthbrooders as foster parents. In the lake S. multipunctatus

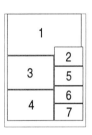

page 248

1. A male *Ctenochromis horei* at Isanga, Zambia.
2. A mouthbrooding female *C. horei* finds shelter among the rocks (Msalaba, Tanzania).
3. A fry-guarding female *C. horei* at Isanga.
4. *Synodontis multipunctatus* in the aquarium. Photo by Mark Smith.
5. A juvenile *S. multipunctatus*, a brood parasite, feeding on a cichlid egg inside the cichlid's mouth. Photo by Max Bjørneskov.
6. *Synodontis dhonti* at Katete, Zambia.
7. *Synodontis nigromaculatus* at Katete. This catfish also occurs outside Lake Tanganyika.

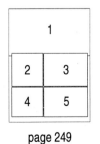

page 249

1. *Limnotilapia dardennii* at Chisanza, Zambia.
2. A male *L. dardennii* at his large nest (Mvuna Island, Tanzania).
3. Juvenile *L. dardennii* feeding on water plants (Isanga, Zambia).
4 and 5. *Simochromis loocki* feeds on algae growing on water plants (Isanga).

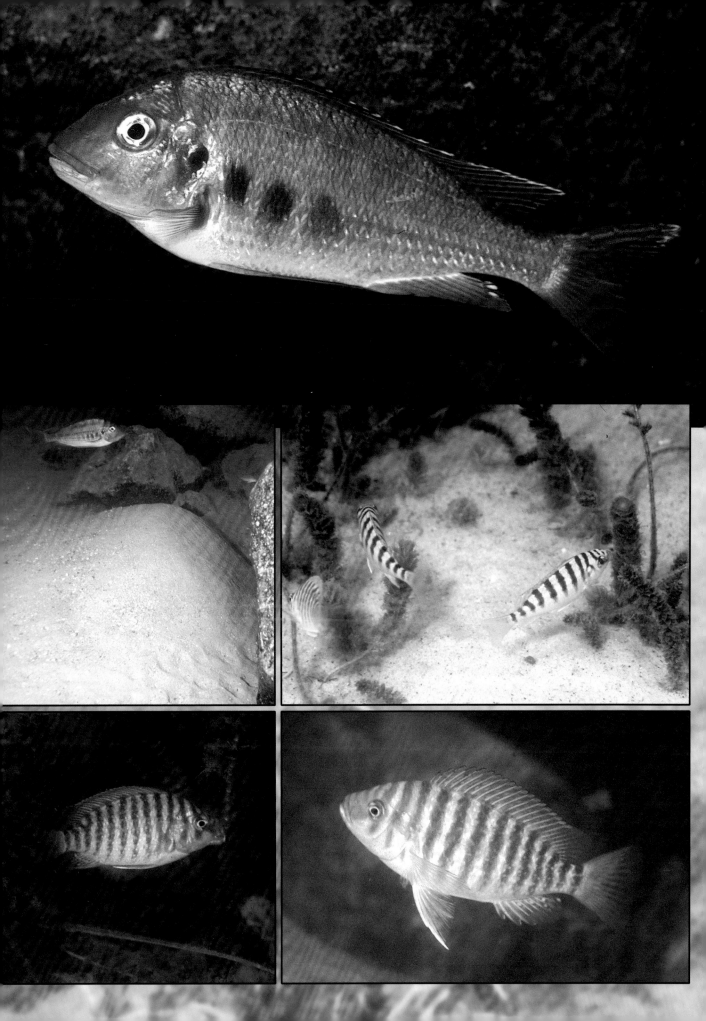

uses several different species as foster parents, with *C. horei* and *Simochromis babaulti* being the most common (Sato, 1986).

Limnotilapia dardennii

L. dardennii has a lake-wide distribution and is rather common in the shallow sediment-rich habitat, in particular near vegetation. It is a large species and valued by fishermen; large numbers can usually be found in local markets. Males can attain a maximum total length of about 26 cm; females are rarely larger than 20 cm. Adult *L. dardennii* are most often found over sandy bottoms in or near vegetation, sometimes far away from rocky substrates; juveniles are very common in the extreme shallows; and territorial males usually construct their nests in the intermediate habitat at somewhat deeper levels — nests are commonly found between depths of 3 and 10 metres. It is an omnivore but with a liking for plants and algae.

Males, which are quite aggressive towards conspecific males in their territorial defence, build huge nests in the intermediate habitat. A large sand mound, with a height of more than 50 cm, is erected very close to a large rock, and the side facing the rock is prepared as the spawning site (see photo page 249). Spawning takes place on the inner side of the nest, sometimes in the narrow space near the bottom between the sand cone and the rock. Because of the female's size clutches can number more than 100 eggs, and the eggs themselves are large as well: 6 mm (Poll, 1956). Mouthbrooding females gather in small groups and idle on the bottom near rocky habitats. Poll also mentions that *L. dardennii* has its peak breeding season from March to May. Breeding males and mouthbrooding females can, however, be seen all year round.

Although the male has a splendid breeding coloration, geographical variation is unknown. This is probably due to the species' habit of crossing large sandy habitats in search of vegetation.

Simochromis

Simochromis species are to the shallow sediment-rich habitat, what *Tropheus* species are to the pure rocky habitat. They have a similar morphology and also feed on algae, which is picked from the rocks. However, in this habitat the rocks are covered with sand and mud, and the algae are found on those stones and pebbles that are kept relatively clean from sediment by the movement of the extremely shallow water. *Simochromis* species collect the fine layer of diatoms and algae covering the substrate and are true vegetarians. Their intestines are three to five times as long as their total length (Poll, 1956), which is indicative of a herbivore.

The two most common *Simochromis* species in the shallow habitat are *S. babaulti* and *S. diagramma*. Both species have a lake-wide distribution, but only *S. babaulti* has evolved geographical variants. The males of the population in the southwestern part of the lake, which has been described as a different species (*S. pleurospilus*) have many rows of red dots on the sides. In northern populations there are only a few red dots on the male's flank. Males are further characterized by a short black stripe in the spiny part of the dorsal fin, a feature not found in any other Tanganyika cichlid. The other sex lacks such a stripe but is adorned with cherry-red patches on the cheeks and anal fin. The maximum total length of *S. babaulti* is about 11 cm whereas that of *S. diagramma* is 20 cm.

S. babaulti occurs in the very shallow water, and is rarely found deeper than two metres. *S. diagramma* is usually found deeper than *S. babaulti* but most often no deeper than four metres. Sometimes large schools of adult *S. diagramma* can be seen roaming around in the habitat. Both species are territorial and stake out their territories in the rocky part of the habitat. Territorial *S. babaulti* males are found very close to the shoreline in water less than one metre deep — probably because of the extremely shallow water, large predators are very rare. Their territories consist of a few small stones that form a rather even platform. The spawning sites of *S. diagramma* are hidden in the few caves of the habitat and males guard the entrances to their caves vigorously against all intruders.

Two other species, *S. marginatus* and *S. margaretae*, occur in the shallow habitat, but their range appears to be restricted to the northern section of the lake. *S. marginatus*, which attains a similar size to *S. babaulti*, was described from a single specimen collected in the Congo along the Ubwari Peninsula, but has also been found near Nyanza Lac (Herrmann, 1992) and at Miyako in Tanzania (Kuwamura, 1987b). Males of this species are characterized by a black marginal band in the dorsal fin; females lack such a band. *S. margaretae* is very similar to *S. marginatus* — males have a black marginal band in the dorsal — and is probably conspecific with the latter. The only difference lies in the depth of the caudal peduncle: that of *S. margaretae* is deeper than long, while that of *S. marginatus* is longer than deep. *S. margaretae* has been found only in Kigoma Bay.

Simochromis loocki

S. loocki, whose taxonomic status is currently under investigation, seems to have a lake-wide range although no records are known from the extreme northern part of the lake. The only place I have seen *S. loocki* is near Isanga in Zambia. These individuals were all juveniles or semi-adult specimens that were feeding mainly from water plants. Just as *Petrochromis* species comb algae from rocks, *S. loocki* combs algae from plants! Yamaoka (1997) mentions that this species — which he calls *Interochromis loocki* — feeds exclusively on unicellular algae, but also that it feeds from rocky substrates and even from muddy bottoms at a depth of 30 metres. It may thus be that the individuals I found in the shallow sediment-rich habitat are juveniles that feed by combing diatoms from plants, but which, when they grow older, move to other habitats. The maximum total length of *S. loocki* is about 10 cm.

Large cichlids

Tilapias are very common in African rivers and small lakes, and are the hardiest members of the cichlid family; they can stand very high water temperatures, very bad water conditions, very poor nutrition, and very hostile habitats. They are all-rounders that can easily adapt not only to adverse situations but also to new habitats. But equally, because they are generalists they can be "outdone" by scores of other cichlids that are specialized in a single particular field and which live in environments that change very little. Such situations are found in Africa's great lakes, and the tilapiines in these lakes, though common, haven't diversified as much as the other cichlids.

Only a single tilapiine occurs in Lake Tanganyika: *Oreochromis tanganicae*. Two other species, *O. niloticus* and *O. karomo*, occur in the swamps surrounding the lake, but have not been found in the lake itself. *O. tanganicae* can attain a total length of more than 40 cm and weigh more than two kilos. It is thus a species much coveted by the local fishermen, who catch it using gill nets. It is commonly found in large foraging schools that roam through different types of habitat, but most often found in sediment-rich environments. It is a herbivore, feeding off the layer of algae and diatoms that covers sand, plants, and rocks; this loose material is collected from the

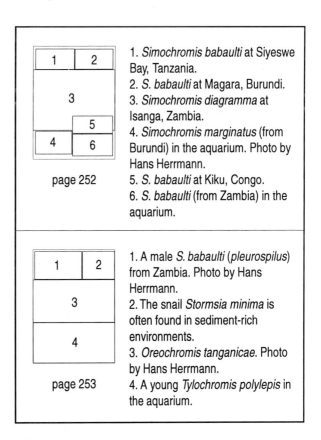

1. *Simochromis babaulti* at Siyeswe Bay, Tanzania.
2. *S. babaulti* at Magara, Burundi.
3. *Simochromis diagramma* at Isanga, Zambia.
4. *Simochromis marginatus* (from Burundi) in the aquarium. Photo by Hans Herrmann.
5. *S. babaulti* at Kiku, Congo.
6. *S. babaulti* (from Zambia) in the aquarium.

page 252

1. A male *S. babaulti* (*pleurospilus*) from Zambia. Photo by Hans Herrmann.
2. The snail *Stormsia minima* is often found in sediment-rich environments.
3. *Oreochromis tanganicae*. Photo by Hans Herrmann.
4. A young *Tylochromis polylepis* in the aquarium.

page 253

biocover by scraping movements of the lips, which are closely set with long and movable teeth, not unlike those of *Petrochromis* species, and then pre-processed by the many teeth in the pharyngeal jaws. The intestinal tract is about six times as long as the fish's total length (Poll, 1956).

O. tanganicae is a mouthbrooder in which males construct very large sand-castle nests, which can be found in very shallow water, near vegetation. Mouthbrooding females gather in large groups and do not eat while holding eggs and larvae.

Another mouthbrooder, *Tylochromis polylepis*, seems to be in a kind of transitional phase between a riverine and a truly lacustrine species. In the lake it occurs only near river mouths, sometimes in large numbers, but it is also found upstream in rivers. Poll (1956) assumes that this species feeds on higher plants such as *Vallisneria*, but the molariform teeth on the pharyngeal bones, and the relatively short intestinal tract (twice the fish's total length), may favour the idea that it is a generalized invertebrate-feeder with the ability to crush small mollusc shells. Stomach contents inventories have been found to include large quantities of sand grains, indicating that food is collected from the bottom. Nothing is known about the breeding behaviour of *T. polylepis,* and it may not even breed in the lake itself.

Muddy bottoms

Muddy bottoms are common in Lake Tanganyika, and most of the bottoms deeper than 60 metres are muddy. The habitat to be discussed here is flat or has a minimal gradient, and is usually found at depths ranging between 10 and 50 metres. The mud is not a very loose type of sediment but, being mixed with fine-grained sand, forms a rather firm substrate which is stable enough to support holes and tunnels dug in it. Some species have developed lifestyles that involve living in and around these muddy bottoms, and there are even two tiny substrate brooders — among the smallest cichlids known from the lake — found in this very open habitat.

Tiny mud-tunnel brooders

Neolamprologus kungweensis and *N. signatus* have in the past suffered some confusion as regards their true identity (Konings, 1988). *N. kungweensis* was described from specimens collected near Bulu Point in Tanzania (Poll, 1952, 1956), all of which were female. At the same time Poll also described *N. signatus*, from specimens collected near Moba in the Congo, and as chance would have it, all those specimens were male. Later (Konings, 1988) it was discovered that *N. kungweensis* from Tanzania exhibits a very distinct sexual dichromatism: females have a very large and ocellated spot in the dorsal fin, but the dorsal fin in males is clear. Morphologically both species are very similar, and, because of the uncommon composition of the type series, an erroneous conclusion was drawn that these two species were conspecific (Konings, 1988). With the subsequent discovery of *N. signatus* in Zambia it became obvious that there are indeed two different species.

Recently *N. laparogramma*, a putative third species of tiny mud-tunnel brooder, has been described (Bills & Ribbink, 1997); the morphology of this form, however, is identical to that of the other two, and in coloration it differs only marginally from *N. signatus*. In fact coloration is the only means of distinguishing the two. On the other hand, even though morphologically *N.*

signatus and *N. kungweensis* are virtually identical, there are a number of features of the colour pattern in both sexes that clearly distinguishes these two forms, plus they have separate distributions in the lake and can be readily identified when encountered underwater. The case of *N. laparogramma* is different: males and females are very similar to those of *N. signatus* and differ only marginally in details of their colour pattern. *N. laparogramma* was described from specimens collected near Mpulungu and Mbita (Mbete) Island, and males differ from *N. signatus* males from Sumbu and Moba only insofar as the latter have up to 13 vertical bars on the body, and the former up to 8, mainly on the posterior part of the body. The difference in coloration between females of the two forms is not apparent, at least not in aquarium specimens, but consists of a different pattern of pigmented scales on the belly.

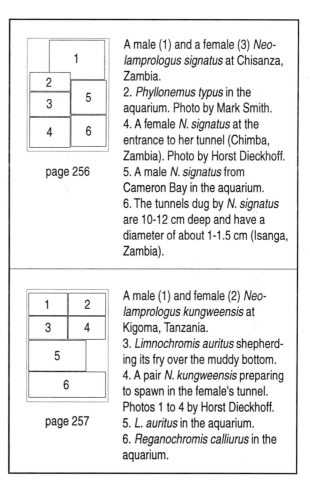

A male (1) and a female (3) *Neolamprologus signatus* at Chisanza, Zambia.
2. *Phyllonemus typus* in the aquarium. Photo by Mark Smith.
4. A female *N. signatus* at the entrance to her tunnel (Chimba, Zambia). Photo by Horst Dieckhoff.
5. A male *N. signatus* from Cameron Bay in the aquarium.
6. The tunnels dug by *N. signatus* are 10-12 cm deep and have a diameter of about 1-1.5 cm (Isanga, Zambia).

page 256

A male (1) and female (2) *Neolamprologus kungweensis* at Kigoma, Tanzania.
3. *Limnochromis auritus* shepherding its fry over the muddy bottom.
4. A pair *N. kungweensis* preparing to spawn in the female's tunnel. Photos 1 to 4 by Horst Dieckhoff.
5. *L. auritus* in the aquarium.
6. *Reganochromis calliurus* in the aquarium.

page 257

In *N. laparogramma* these scales form 5 to 9 vertical lines on the belly whereas in *N. signatus* they form an irregular pattern.

Up to now geographical variants distinguished solely by a different colour pattern have not generally been regarded as different species, and there is, in my opinion, equally no reason to treat these two colour forms of *N. signatus* any differently. No difference other than coloration has been found, and I therefore suggest that *N. laparogramma* should be regarded as a junior synonym of *N. signatus*.

N. signatus and *N. kungweensis* are found mainly over flat muddy bottoms, and their presence is given away by numerous holes in the bottom. These holes have a diameter of about 1 cm and are about 12 cm deep. Casts of such living quarters have shown that they are not dug straight down but always have an angle (Bills & Ribbink, 1997). Male and female have their own holes and usually one sees more holes than occupants. The holes are usually about 50 cm apart.

The diet of these little cichlids consists of zooplankton and other small invertebrates that are encountered while digging, and maybe any that try to shelter in the fish's hole. Such holes form ideal traps for anything that crawls around the bottom.

Bills & Ribbink mention that these mud-brooders are monogamous and that territories include holes for the female as well as for the male. Spawning takes place inside the female's hole and, remarkably, the eggs of this substrate brooder do not stick! It is known from aquarium observations that *N. signatus* eggs laid in an empty shell do not stick to the inside of the shell but rather roll about whenever the shell is picked up. Büscher (1998) reports that the eggs of *N. kungweensis* are likewise non-adhesive. Interestingly, it has long been known that *Hypsophrys nicaraguensis*, a substrate spawning cichlid from Central America, also lays non-adhesive eggs, and more recently it has been found that this species too digs holes, in the mud-sand substrate of vertical river banks (Ron Coleman, pers. comm.). It thus appears that in at least some cases there is a correlation between cichlids spawning in holes with crumbly walls and the production of eggs and larvae that don't adhere to the substrate.

It is obvious that this is beneficial to the eggs because a (partial) collapse of the hole will not bury and suffocate the much lighter eggs under a load of mud and sand.

One could even argue that non-adhesive eggs, adapted for a mud-hole lifestyle, may be at the root of some instances of mouthbrooding: parents can easily (re)move eggs or larvae from a collapsed or damaged hole to a better one or, alternatively, keep them inside their mouths till they are independent.

Of course, the concept of mouthbrooding was not developed overnight, and this mechanism for protecting the offspring in their initial stages of life is thought to have developed several times among the different lines of Tanganyika cichlids, probably in different circumstances and with different triggers; so maybe on at least one such occasion the loose eggs of a substrate brooder may have been a step in the right direction.

A mouthbrooding mud-hole dweller

Limnochromis auritus is a mud-dweller and a mouthbrooder, and its forbearers may have been substrate-brooding mud-dwellers that made the switch to mouthbrooding. It is a biparental mouthbrooder with small and numerous eggs and the brooding period is also very short. In this respect it is not much removed from the substrate brooders.

L. auritus, whose maximum total length is about 17 cm, is found throughout the lake and is recorded from great depths as well as from shallow water. It does not like bright light and hides in tunnels and holes which are dug in the muddy bottom. At deeper levels, where daylight hardly penetrates, it may live more in the open. *L. auritus* is a schooling species which digs numerous tunnels several feet deep (Horst Dieckhoff, pers. comm.). These tunnels have underground connections, perhaps so that the fish can escape the attentions of an inquisitive water cobra (ditto fish collectors!).

L. auritus feeds on snails and other invertebrates that are probably uncovered during its digging activities. Juveniles feed on zooplankton. The intestinal tract is only 30% of the fish's total length, indicating a rapidly digestible food.

Spawning takes place inside a tunnel, and although male and female may share a single tunnel, aquarium observations suggest that a very narrow tunnel, wide enough for only a single fish, is preferred for spawning. In the aquarium spawning is preceded by intensive digging and shovelling, moving large quantities of sand which is piled up against the front glass creating a big hole at the rear of the aquarium. If enough sand is provided to screen the pair from the sight of the outside world, spawning may follow. It is also important that some form of dark cave is available for the pair to spawn in — the best option is a plastic pipe with a diameter of around 5 cm and a length of about 50 cm (Mackie Kilts, pers. comm.). One end of the pipe should be closed.

Spawning begins with the male entering the tunnel and discharging his seminal fluid. After the male has backed out of the tunnel the female enters and starts laying eggs. The eggs are probably fertilized as soon as they are expelled by the female. Numerous eggs are produced, and these are gathered up by the female, and sometimes also by the male if he has direct access to them, though he subsequently spits "his" eggs out in front of the female. Every now and then the female interrupts oviposition and backs out of the tunnel. She then swims towards the male and nuzzles his vent region; this probably induces him to enter the tunnel and release some more milt. After the male has again retreated from the tunnel the female enters again and continues laying eggs (Baasch, 1987; Kilts, pers. comm.). This spawning technique is similar to that found in *Cyphotilapia frontosa* (albeit in a tunnel), where oviposition is likewise preceded by the male discharging seminal fluid on the spawning site.

After a while too many eggs will have been expelled to be comfortably fitted into the buccal cavity of the female. At this point the male collects the eggs and this time retains them in his mouth so that now both parents are brooding. After about two days either the number of eggs has decreased or the eggs/larvae will now fit inside a single parent's mouth for some other reason, as now only the male or female broods at any one time.

The larvae are frequently exchanged between the parents, allowing each to feed in turn. During such exchanges the complete brood, which can number hundreds, is spat out in front of the other parent, who then takes up the larvae and

continues brooding. As mouthbrooders go, the incubation period is very short and the fry are released after nine days. Both parents guard their progeny for rather a long period after they have been released.

The spawning technique of *L. auritus* in some ways resembles that of a substrate brooder, i.e. the female deposits a row of eggs which are fertilized by the male's semen as soon as they are expelled. The male remains at a distance while the female lays eggs. The difference from true substrate brooders is that the eggs have not yet been laid when the male discharges his milt to fertilize them.

Two other species of the same genus, *L. staneri* and *L. abeelei*, may breed in a similar fashion, but because they are found in very deep water nothing is known of their lifestyles (see page 134).

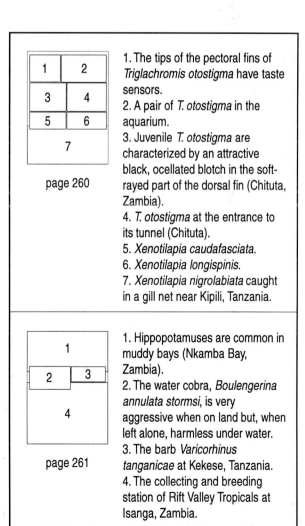

1. The tips of the pectoral fins of *Triglachromis otostigma* have taste sensors.
2. A pair of *T. otostigma* in the aquarium.
3. Juvenile *T. otostigma* are characterized by an attractive black, ocellated blotch in the soft-rayed part of the dorsal fin (Chituta, Zambia).
4. *T. otostigma* at the entrance to its tunnel (Chituta).
5. *Xenotilapia caudafasciata*.
6. *Xenotilapia longispinis*.
7. *Xenotilapia nigrolabiata* caught in a gill net near Kipili, Tanzania.

1. Hippopotamuses are common in muddy bays (Nkamba Bay, Zambia).
2. The water cobra, *Boulengerina annulata stormsi*, is very aggressive when on land but, when left alone, harmless under water.
3. The barb *Varicorhinus tanganicae* at Kekese, Tanzania.
4. The collecting and breeding station of Rift Valley Tropicals at Isanga, Zambia.

The mud-eater

Triglachromis otostigma is a common mouth-brooding cichlid, found in suitable muddy habitat throughout the lake. It is well adapted for life on muddy floors, one special adaptation being its peculiar pectoral fins. The lower seven to eight rays of the pectoral fins have free tips that are not attached to a membrane, and which can be bent downwards towards the bottom to act as sensory instruments. In captivity the pectorals come into action after food has been put into the aquarium; when food is visually located it is taken from the water column, but if *T. otostigma* smells food but doesn't see any it begins screening the bottom of the tank with its pectoral fins.

T. otostigma is the only cichlid known so far from the lake that feeds on mud (Coulter, 1991). It palpates the muddy sediment with the filamentous extensions of its pectoral fin rays, and when something interesting has been located it moves backward and then scoops up mouthfuls of mud, hopefully including the item it sensed previously. Specimens in the wild have been found with guts full of mud (Coulter, 1991). The intestinal tract measures 2.5 times the fish's total length (maximum 10 cm) suggesting a diet rich in "ballast" material (such as mud).

In its natural environment *T. otostigma* digs large holes and tunnels in the soft bottom and is found in shallow water near river mouths as well as at deep levels. Usually there is a complex of several holes close together, at a distance of less than 50 cm from each other. The distance from one cluster to another is several metres. It seems that each cluster is inhabited by a family consisting of a male, a female, and several juveniles. The entrance to a hole is quite large and has a diameter of about 5 cm and is sometimes marked by a surrounding ring of sand.

In captivity breeding in *T. otostigma* is preceded by vigorous and extensive digging. The gravel in most tanks is unsuitable for tunnel digging, and instead large pits are excavated. Breeding in captivity appears to be enhanced by providing a pair with ample material to dig. Furthermore a dark tunnel (PVC pipe with a diameter of about 5 cm) seems indispensable. The small eggs — Poll (1956) records up to 100 eggs, with a size of no more than 1 mm, in a ripe female — are deposited inside the cave. The pair stays together and the fry are probably exchanged from parent to parent several times during incubation. The juveniles stay within their parents' burrow, and even at a size of about 4 cm they can be seen going in and out of the large holes. The juveniles have a very characteristic ocellated spot in the dorsal fin (it looks almost like a so-called *Tilapia*-spot), which disappears when they reach maturity.

Reganochromis calliurus

R. calliurus has a breeding technique similar to that of *L. auritus*. The difference is that *R. calliurus*, which normally lives over muddy bottoms, does not dig tunnels or large pits (at least not in captivity) but prefers a dark cave among the rocks in which to spawn. Although *R. calliurus* lives over the mud floor it has never been found digging for food in the aquarium. It feeds on shrimps, tiny crabs, and small fishes (Poll, 1956), which it scoops from the bottom. Maximum total length is about 15 cm; females remain a bit smaller than males. *R. calliurus* is common throughout the lake but never found in great numbers.

Dickmann (1986) reported that broods can number up to 60 fry. The eggs are considerably larger (2.5 mm) than those of *L. auritus* (1 to 1.5 mm). The larvae are frequently exchanged between the parents and incubation is rather short.

Xenotilapia

Four species of *Xenotilapia* are thought to live on muddy bottoms, but none of them have been observed in their natural habitat. Some of them simply live too deep to be observed in the wild, while others may live in water too turbid for visual observation.

X. nigrolabiata, a species recorded only from deep water, grows to a maximum total length of 13 cm, and males in breeding colour are characterized by a black upper lip. Nothing is known about their breeding behaviour, but since the male has a very conspicuous breeding dress it probably belongs to the group of maternal mouthbrooders. *X. nigrolabiata* feeds on zooplankton (Poll, 1956).

Two other species in this habitat, *X. longispinis* and *X. burtoni*, feed on the "normal" *Xenotilapia* diet of insect larvae and crustaceans. These two species are characterized by an enlarged spiny section to the dorsal fin, which is obvious in *X. longispinis* but very prominent in *X. burtoni*. The anterior part of the dorsal fin is marked with a number of black patches in both species, but the black margin to the tail, seen in *X. longispinis*, is absent in *X. burtoni*. *X. burtoni* is one of the very few species of the genus found on sandy or muddy bottoms which has a limited distribution: it has been caught only in Burton Bay. In my opinion *X. burtoni* is very likely conspecific with *X. longispinis* and just a geographical variant of the latter species. But without having seen either species in the wild, such suggestions are, of course, largely without value.

X. caudafasciata is found throughout the lake and has been caught at relatively deep levels, mostly below 40 metres (Poll, 1956). Males are beautifully marked, with two or three vertical black bars on the tail and a horizontal bar, perpendicular to the rays, on the pelvic fins. A more surprising characteristic is the fact that all the males captured have been on average 1.5 cm smaller than females. Males attain an average length of about 10.5 cm whereas females attain, on average, 12 cm, and have a recorded maximum total length of 15.6 cm. Another feature that sets *X. caudafasciata* apart from other species of *Xenotilapia* is the position of the outer teeth in the lower jaw. In all other *Xenotilapia* species these teeth point outward in an almost horizontal plane. It has been suggested that teeth thus positioned might act like the prongs of a fork and facilitate digging into the sand. The outer teeth in *X. caudafasciata*, however, point in quite the opposite direction: inward. This may be a feeding adaptation (e.g. for coping with ooze that is filtered for prey), or else may indicate that this species does not share the same ancestors as the other *Xenotilapia* species.

Afterword

Although our knowledge of Lake Tanganyika cichlids has developed apace in recent years much remains to be learned, many puzzles to be solved, numerous questions to be answered. Over the past decade or so, there has been an increasing realisation by the scientific world that study of cichlids in their natural environment is an important, even essential, supplement to traditional, morphometric, taxonomic study. And, moreover, that serious aquarists, with a keen eye for detail and readiness to keep meticulous records, are in a position to make a valuable contribution to our knowledge of fishes and their behaviour. Cooperation by all interested parties can only prove beneficial. It is thus my hope that this book will prove of value to scientists interested in these cichlids in their natural environment, and that it will also enable aquarists to create an appropriate setting for their charges. Perhaps a few (from both groups!) may even be inspired by it to follow in my footsteps and visit Lake Tanganyika, to see for themselves its wonderful cichlids, in their natural habitat.

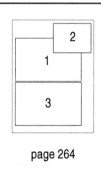

page 264

1. and 2. The fish collectors of Rift Valley Tropicals travel long distances to catch particular cichlids. After a trip lasting several days they have to catch as many as possible and bring the catch back alive to the station. The sexes are separated in order to minimize losses.
3. Preparing to camp for the night at Wampembe, Tanzania.

page 265

A party of visitors from America at the station of Rift Valley Tropicals displaying their underwater equipment. The owner of the station, Toby Veall (1), assisted with the logistics of several of the author's trips. The author, Ad Konings (5), is accompanied by Laif DeMason (4) and the Somermeyers, Steve (3) and his wife Kimberley (2). Photo courtesy of Steve Somermeyer.

References

ALLEN, B. (1996) Spawning *Bathybates minor* — an aquarium first. *Cichlid News*, 5 (2): 22-24

BAASCH, P. (1987) Maulbrüter mit Elternfamilie: *Limnochromis auritus. DCG-Info*, 18 (4): 66-71.

BAILEY, R.M. & D.J. STEWART (1977) Cichlid fishes from Lake Tanganyika: additions to the Zambian fauna including two new species. *Occ. Papers Mus. Zool.*, University of Michigan, 679: 1-30.

BARLOW, G.W. (1991) Mating systems among cichlid fishes. In: M.H.A. Keenleyside (Ed.) *Cichlid Fishes. Behaviour, ecology and evolution.* Chapman & Hall, London.

BILLS, R. & A. RIBBINK (1997) Description of *Lamprologus laparogramma* sp. nov., and re-diagnosis of *Lamprologus signatus* Poll 1956 and *Lamprologus kungweensis* Poll 1952, with notes on their ecology and behaviour (Teleostei; Cichlidae). *S. Afr. J. Sci.*, 93: 555-565.

BOULENGER, G.A. (1898) Report on the collection of fishes made by Mr. J.E.S. Moore in Tanganyika during his expedition 1895-1896. *Trans. Zool. Soc. Lond.* XV. prt 1, nr. 1: 1-30.

BOULENGER, G.A. (1899) Second contribution to the ichthyology of Lake Tanganyika. On the fishes obtained by the Congo Free State expedition under Lieut. Lemaire in 1898. *Trans. Zool. Soc. Lond.* XV, part 9: 87-95.

BOULENGER, G. A. (1900) Diagnoses of new fishes discovered by Mr. J.E.S. Moore in Lake Tanganyika. *Ann. Mag. N.H.* (7) 7: 478-481.

BRICHARD, M. (1979) The "Sunset" — a new fish from Lake Tanganyika. *TFH*, 27 (#276):40-42.

BRICHARD, P. (1978) *Fishes of Lake Tanganyika.* TFH Publ., Neptune, New Jersey, USA.

BRICHARD, P. (1989) *Book of cichlids and all the other fishes of Lake Tanganyika.* TFH Publ., Neptune, New Jersey, USA.

BÜSCHER, H.H. (1989) Ein neuer Tanganjika-Cichlide aus Zaire — *Neolamprologus marunguensis* n. sp. (Cichlidae; Lamprologini). *DATZ*, 42: 739-743.

BÜSCHER, H.H. (1991a) Neuen Schneckencichliden aus dem Tanganjikasee — *Lamprologus meleagris* n. sp. und *L. speciosus* n. sp. (Cichlidae; Lamprologini) *DATZ*, 44: 374-382.

BÜSCHER, H.H. (1991b) Ein neuer Tanganjika-Cichlide aus Zaire — *Neolamprologus pectoralis* n. sp. (Cichlidae; Lamprologini). *DATZ*, 44: 788-792.

BÜSCHER, H.H. (1992a) *Neolamprologus leloupi* und *N. caudopunctatus. DATZ*, 45: 39-44.

BÜSCHER, H.H. (1992b) Verbreitung und Ökologie von *Neolamprologus buescheri. DATZ*, 45: 305-310.

BÜSCHER, H.H. (1992c) Ein neuer Cichlide aus dem Tanganjikasee — *Neolamprologus similis* n. sp. *DATZ*, 45: 520-525.

BÜSCHER, H.H. (1992d) *Neolamprologus nigriventris* n. sp.: ein neuer Tanganjikasee-Cichlide (Cichlidae, Lamprologini). *DATZ*, 45: 778-783.

BÜSCHER, H.H. (1997) Ein neuer Cichlide aus dem Tanganjikasee — *Neolamprologus helianthus* n. sp. (Cichlidae, Lamprologini). *DATZ*, 50: 701-706.

BÜSCHER, H.H. (1998) Buntbarsche in Schneckenhäusern. *DATZ Sonderheft Tanganjikasee*, 6: 51-59.

CAPART, A. (1949) Sondages et carte bathymétrique du Lac Tanganika. Résultats scientifiques de l'exploration hydrobiologique du Lac Tanganika (1946-1947). *Inst. royal Sci. Nat. Belg.*, 2 (2): 1-16.

COHEN, A.S., M.G. SOREGHAN & C.A. SCHOLTZ (1993) Estimating the age of formation of lakes: an example from Lake Tanganyika, East-African Rift system. *Geology*, 21: 511-514.

COLOMBE, J. & ALLGAYER, R. (1985) Description de *Variabilichromis, Neolamprologus* et *Paleolamprologus*, genres nouveaux du lac Tanganyika, avec redescription des genres *Lamprologus* Schilthuis 1891 et *Lepidiolamprologus* Pellegrin 1904. *Rev. Franc. des Cichlidophiles*, 49: 9-28.

COULTER, G.W. (1991) *Lake Tanganyika and its life.* Oxford University Press; London & New York.

DeVOS, L., M. NSHOMBO, & D. THYS van den AUDENAERDE (1996) *Trematocara zebra* (Perciformes: Cichlidae), nouvelle espèce du nord-ouest du lac Tanganyika (Zaire). *Belg. J. Zool.* 126: 3-20.

DeVOS. L., L. RISCH, & D. THYS van den AUDENAERDE (1996) *Xenotilapia nasus*, nouvelle espèce de poisson des zones sous-littorale et benthique du nord du lac Tanganyika (Perciformes: Cichlidae). *Ichthol. Explor. Freshwaters*, 6 (4): 377-384.

DICKMANN, H-B. (1986) *Reganochromis calliurum.* Der Sanfte aus dem Tanganjikasee. Bericht über erste (?) Nachzuchten. *Das Aquarium* 206 (8): 402-406.

ECCLES, D.H. & E. TREWAVAS (1989) *Malawian cichlid fishes — The classification of some Haplochromine genera.* Lake Fish Movies, Herten, Germany.

EYSEL, W. (1990) Tanganjikasee-Cichliden: Die Übergattung Ectodini — 1. *Ectodus descampsi. DCG-Info*, 21 (6): 119-124.

EYSEL, W. (1992) Staubsauger-Cichlide. Vorkommen, Pflege und Vermehrung von *Gnathochromis permaxillaris*. *Das Aquarium*, 274: 12-17.

FOHRMAN, K. (1994) Bred in an aquarium: *Boulengerochromis microlepis*. *Cichlids Yearbook*, 4: 22-23.

FRYER, G. & ILES. T.D. (1972) *The cichlid fishes of the Great Lakes of Africa*. TFH Publ., Neptune, New Jersey, USA.

GASHAGAZA, M.M. (1991) Diversity of breeding habits in Lamprologini cichlids in Lake Tanganyika and the mechanism of their coexistence. *Physiol. Ecol. Japan*, 28: 29-65.

GASHAGAZA, M.M., K. NAKAYA & T. SATO (1995) Taxonomy of small-sized cichlid fishes in the shell-bed area of Lake Tanganyika. *Jap. J. Ichthyol.*, 42: 291-302.

GEERTS, M. (1991) The last minutes of speciation. *Cichlids Yearbook*, 1: 94-95.

GREENWOOD, P.H. (1965) The cichlid fishes of Lake Nabugabo, Uganda. *Bull. Brit. Mus (N. Hist.) Zool*. 12 (9): 315-357.

GREENWOOD, P.H. (1983) The *Ophthalmotilapia* assemblage of cichlid fishes reconsidered. *Bull. Br. Mus. Nat. Hist. (Zool.)*, 44: 249-Z90.

GREENWOOD, P.H. (1984) What is a Species Flock? In: A.A. Echelle & I. Kornfield (Eds). *Evolution of fish species flocks*. Univ. Maine at Orono Press. USA.

HABERYAN, K.A. & HECKY, R.E. (1987) The late Pleistocene and Holocene stratigraphy and paleolimnology of Lakes Kivu and Tanganyika. *Paleogeography, Paleoclimatology, Paleoecology*. 61: 169-197.

HERRMANN, H-J. (1990) *Die Buntbarsche der Alten Welt. Tanganjikasee*. Ulmer Verlag, Stuttgart, Germany.

HERRMANN, H.-J. (1992) *Simochromis marginatus*. *Cichlids Yearbook*, 2: 16.

HERRMANN, H.J. (1994a) *Ophthalmotilapia nasuta* from Resha, Burundi. *Cichlids Yearbook*, 4: 12-13.

HERRMANN, H.J. (1994b) An intriguing yellow *Petrochromis*. *Cichlids Yearbook*, 4: 14-17.

HERRMANN (1995) Breeding in *Lamprologus meleagris* Büscher, 1991. *Cichlids Yearbook*, 5: 18-19.

HERRMANN, H.J. (1996) On the differences between *Petrochromis polyodon* and *P. famula*. *Cichlids Yearbook*, 6: 18-19.

HORI, M. (1983) Feeding ecology of thirteen species of *Lamprologus* (Teleostei: Cichlidae) coexisting at a rocky shore of Lake Tanganyika. *Physiol. Ecol. Japan*, 20: 129-149.

HORI, M. (1987) Mutualism and commensalism in a fish community in Lake Tanganyika. Pp 219-239 in: *Evolution and coadaptation in biotic communities* (S. Kawano, J.H. Connell & T. Hidaka; Eds.) Univ. Tokyo Press, Tokyo.

HORI, M. (1993) Frequency-dependent natural selection in the handedness of scale-eating cichlid fish. *Science*, 260: 216-219.

KARLSSON, M. (1998) Från tidens vägskäl. *Ciklidbladet*, 4/98: 22-37.

KASSELMANN (1998) Wasserpflanzen. *DATZ Sonderheft Tanganjikasee*, 6: 26-29.

KAWANABE, H. (1981) Territorial behaviour of *Tropheus moorei* (Osteichthyes: Cichlidae) with a preliminary consideration on the territorial forms in animals. *Afr. Stud. Monogr.*, 1: 101- 108.

KOCHER, T.D., J.A. CONROY, K.R. McKAYE, J.R. STAUFFER & S.F. LOCKWOOD (1995) Evolution of NADH dehydrogenase subunit 2 in East African cichlid fish. *Mol. Phylogenet. Evol.*, 4: 420- 432.

KOHDA, M. & M. HORI (1993) Dichromatism in relation to the tropic biology of predatory cichlid fishes in Lake Tanganyika, East Africa. *J. Zool.*, 229: 447-455.

KOHDA, M. & Y. YANAGISAWA (1992). Vertical distributions of two herbivorous cichlid fishes of the genus *Tropheus* in Lake Tanganyika, Africa. *Ecol. Freshw. Fish*, 1: 99-103.

KONINGS, A. (1980) Aquarist's guide to *Lamprologus brevis*. *Buntbarsch Bulletin*, 77: 3-7.

KONINGS, A. (1988) *Tanganyika Cichlids*. Verduijn Cichlids. Zevenhuizen, Holland.

KONINGS, A. (1991a) *Neolamprologus* sp. "Cygnus". *Cichlids Yearbook*, 1: 11.

KONINGS, A. (1991b) *Microdontochromis tenuidentatus*. *Cichlids Yearbook*, 1: 19.

KONINGS, A. (1992) Clues to a step-wise speciation. *Cichlids Yearbook*, 2: 6-9.

KONINGS, A. (1993) The *Neolamprologus brichardi* complex. *Cichlids Yearbook*, 3: 6-13.

KONINGS, A. (1995) *Neolamprologus meeli*, an interesting shell-brooder. *Cichlids Yearbook*, 5: 20-21.

KONINGS, A. (1998) A visit to the central Tanzanian coast of Lake Tanganyika. *Cichlid News*, 7 (2): 6-15.

KONINGS, A. & H.W. DIECKHOFF (1992) *Tanganyika secrets*. Cichlid Press, St. Leon-Rot, Germany.

KRÜTER, R. (1991) *Trematocara nigrifrons* Boulenger, 1906. *Cichlids Yearbook*, 1: 18.

KUWAMURA, T. (1986a) Parental care and mating systems of cichlid fishes in Lake Tanganyika: a preliminary field survey. *J. Ethol.* 4, pp 129-146.

KUWAMURA, T. (1986b) Substratum spawning and biparental guarding of the Tanganyikan cichlid *Boulengerochromis microlepis*, with notes on its life history. *Physiol. Ecol. Japan*, 23 ,pp 31-43.

KUWAMURA, T. (1987a) Male mating territory and sneaking in a maternal mouthbrooder, *Pseudosimochromis curvifrons* (Pisces; Cichlidae). *J. Ethol.*, 5: 203-206.

KUWAMURA, T. (1987b) Distribution of fishes in relation to the depth and substrate at Myako, east-middle coast of Lake Tanganyika. *Afr. Stud. Monogr.*, 7: 1-14.

KUWAMURA, T. (1988) Biparental mouthbrooding and guarding in a Tanganyikan cichlid *Haplotaxodon microlepis. Jap. J. Ichthyol.*, 35: 62-68.

KUWAMURA, T. (1992) Overlapping territories of *Pseudosimochromis curvifrons* males and other herbivorous cichlid fishes in Lake Tanganyika, *Ecol. Res.*, 7: 43-53.

KUWAMURA, T. (1997) The evolution of parental care and mating systems among Tanganyikan cichlids. Pp. 57-86 in: H. Kawanabe, M. Hori, & M. Nagoshi (Eds.). Fish communities in Lake Tanganyika. Kyoto Univ. Press, Kyoto, Japan.

KUWAMURA, T., M. NAGOSHI & T. SATO (1989) Female-to-male shift of mouthbrooding in a cichlid fish, *Tanganicodus irsacae*, with notes on breeding habits of two related species in Lake Tanganyika. *Env. Biol. Fish.*, 24: 187-198.

LIEM, K.F. & D.J. STEWART (1976) Evolution of the scale-eating cichlid fishes of Lake Tanganyika: A generic revision with a description of a new species. *Bull. Mus. Comp. Zool. Hare.*, 147: 319-350.

MATTHES, H. (1959a) Une sous-espèce nouvelle de *Lamprologus leleupi*: *L. leleupi melas. Fol. Sci. Afr. Centr.*, V (1): 18.

MATTHES, H. (1959b) Un Cichlidae nouveau du lac Tanganika — *Julidochromis transcriptus* n. sp. *Rev. Zool. Bot. Afr.*, LX, 1-2: 126-130.

MATTHES, H. (1962) Poissons nouveaux ou interessants du Lac Tanganika et du Ruanda. *Musee Royal de l'Afrique centrale, Tervuren, Sciences Zoologiques*, (8)3: 27-88.

McKAY, K.R. & W.N. GRAY (1984) Extrinsic barriers to gene flow in rock-dwelling cichlids of Lake Malawi. In: A.A. Echelle & I. Kornfield (Eds). *Evolution of fish species flocks*. Univ. Maine, Orono Press.

NAGOSHI, M. 1987. Survival of broods under parental care and parental roles of the cichlid fish, *Lamprologus toae*, in Lake Tanganyika. *Japan. J. Ichthyol.*, 34: 71-75.

NELISSEN, M. (1977) Description of *Tropheus moorii kasabae* n. ssp. (Pisces Cichlidae) from the south of Lake Tanganyika. *Rev. Zool. Afr.*, 91, nr 1: 237-242.

NELISSEN, M. & D. THYS van den AUDENAERDE (1975) Description of *Tropheus brichardi* sp. n. from Lake Tanganyika (Pisces Cichlidae). *Rev. Zool. Afr.*, 89 (4): 974-980.

NISHIDA, M. (1997) Phylogenetic relationships and evolution of Tanganyikan cichlids: a molecular perspective. Pp. 1-24 in: H. Kawanabe, M. Hori, & M. Nagoshi (Eds.). *Fish communities in Lake Tanganyika*. Kyoto Univ. Press, Kyoto, Japan.

NSHOMBO, M., Y. TANAGISAWA & M. NAGOSHI (1985) Scale-eating in *Perissodus microlepis* (cichlidae) and change of its food habits with growth. *Jap. J. Ichthyology*, 32 (1): 66-73.

POLL, M. (1946) Revision de la Faune ichthyologique du lac Tanganika. *Ann. Mus. Congo Belge, Sci. Zool.*, ser I, 4 (3): 141-364.

POLL, M., 1948. Descriptions de cichlidae nouveaux recueillis par la missien hydrobiologique belge au lac Tanganika (1946-1947). *Bull. Mus. R. Hist. Natur. Belgique* XXIV (26): 1-31.

POLL, M. (1956) Poissons Cichlidae, *Résult. scient. Explor. hydrobiol. belge Lac Tanganika* (1946-1947). vol. III (5B): 1-629.

POLL, M. (1974) Contribution a la faune ichthyologique du lac Tanganika d'après les recoltes de P. Brichard. *Rev. Zool. Afr.* 88 (1): 99-110.

POLL, M. (1981) Contribution a la faune ichthyologique du lac Tanganika. Revision du genre *Limnochromis* Regan 1920. Description de trois genres et d'une espece nouvelle: *Cyprichromis brieni. Ann. Soc. R. Zool. Belg.*, III (1- 4): 163-179.

POLL, M. (1984) *Haplotaxodon melanoides* sp. n. du lac Tanganika. *Rev. Zool. Afr.* 98 (3): 677-683.

POLL, M. (1986) Classification des cichlidae du lac Tanganyika tribus, genres et espèces. Mémoires de la classe des sciences. *Académie royale de Belgique. Collection in-8°-2e série, T. XLV*, (2): 1-163.

POLL, M. & H. MATTHES (1962) Trois poissons remarquables du lac Tanganika. *Ann. Mus. royal Afr. Centr. Sci. Zool. 8°* (111): 1-26.

SALVAGIANI, P. (1996) Problems with breeding *Benthochromis tricoti. Cichlids Yearbook*, 6: 15-17.

SATO, T. (1986) A brood parasitic catfish of mouthbrooding cichlid fishes in Lake Tanganyika. *Nature*, 323: 58-59.

SATO, T. & M.M. GASHAGAZA (1997) Shell-brooding cichlid fishes of Lake Tanganyika: their habitats and mating systems. Pp 219-240 in: H. Kawanabe, M. Hori, & M. Nagoshi (Eds.). *Fish communities in Lake Tanganyika*. Kyoto Univ. Press, Kyoto, Japan.

SCHOLZ, C.A. & B.R. ROSENDAHL (1988) Low lake Stands in Lake Malawi and Tanganyika, East Africa, delineated with Multifold seismic Data. *Science*, 240: 1645-1648.

SEEGERS, L. (1992) Neu aus Tansania: Malagarasi-Grundelbuntbarsche. *Aquarium Heute*, 10: 112-117.

STAECK, W., 1980. Ein neuer Cichlide vom Ostufer des Tanganyikasee: *Lamprologus leleupi longior* n. ssp. *Rev. Zool. Afr.*, 94 (1): 11-14.

STIASSNY, M.L.J. (1997) A phylogenetic overview of

the lamprologine cichlids of Africa (Teleostei, Cichlidae): a morphological perspective. *S. Afr. J. Sci.*, 93: 513-523.

STURMBAUER, C. & A. MEYER (1993) Mitochondrial phylogeny of the endemic mouthbrooding lineages of cichlid fishes from Lake Tanganyika in eastern Africa. *Mol. Biol. Evol.*, 10: 751-768.

STURMBAUER, C., E. VERHEYEN & A. MEYER (1994) Mitochondrial phylogeny of the Lamprologini, the major substrate spawning lineage of cichlid fishes from Lake Tanganyika in Eastern Africa. *Mol. Biol. Evol.*, 11: 691-703.

TAKAHASHI, K., Y. TERAI, M. NISHIDA, & N. OKADA (1998) A novel family of short interspersed repetitive elements (SINEs) from cichlids: the pattern of insertion of SINEs at orthologous loci support the proposed monophyly of four major groups of cichlid fishes in Lake Tanganyika. *Mol. Biol. Evol.*, 15(4): 391-407.

TAKAHASHI, T. & K. NAKAYA (1997) A taxonomic review of *Xenotilapia sima* and *X. boulengeri* (Cichlidae; Perciformes) from Lake Tanganyika. *Ichthyol. Res.*, 44 (4): 335-346.

TAKAMURA, K. (1984) Interspecific relationships of aufwuchs-eating fishes in Lake Tanganyika. *Env. Biol. Fish.*, 10: 225-241.

TIERCELIN, J-J. & A. MONDEGUER (1991) The geology of the Tanganyika trough. Pp 7-48 in: G.W. Coulter (Ed.). *Lake Tanganyika and its life*. Oxford University Press; London & New York.

TIJSSELING, G. & I. TIJSSELING (1982) Gedrag van *Tropheus moorii* in het aquarium. *NVC periodiek*, 42: 11-22.

TREWAVAS, E. & M. POLL (1952) Three new species and two new subspecies of the genus *Lamprologus*. Cichlid fishes of the Lake Tanganyika. *Bull. Inst. R. Sci. Natur. Belgique*, XXVIII, 50.1-16.

VAILLANT, L. (1899) *Protopterus retropinnis* et *Ectodus foae*, espèces nouvelles de l'Afrique Equatoriale. *Bull. Mus. Paris*: 219-221.

VERHEYEN, E., L. RÜBER, J. SNOEKS & A. MEYER (1996) Mitochondrial phylogeography of rock-dwelling cichlid fishes reveals evolutionary influence of historical lake level fluctuations of Lake Tanganyika, Africa. *Phil. Trans. Roy. Soc. Lond. B*, 351: 797-805.

WALTER, B. & F. TRILLMICH (1994) Female aggression and male peace-keeping in a cichlid fish harem: conflict between and within the sexes in *Lamprologus ocellatus*. Behav. Ecol. Sociobiol., 34: 105-112.

YAMAOKA, K. (1983a) Feeding behaviour and dental morphology of algae scraping cichlids (Pisces: Teleostei) in Lake Tanganyika. *Afr. Stud. Monogr.*, 4: 77-89.

YAMAOKA, K. (1983b) A revision of the cichlid fish genus *Petrochromis* from Lake Tanganyika, with description of a new species. *Jap. J. Ichthyol.*, 30: 129-141.

YAMAOKA, K. (1990) Feeding behaviour of *Asprotilapia leptura*, an epilithic algal feeding cichlid fish in Lake Tanganyika. *Japan. J. Ichthyol.*, 37: 80-82.

YAMAOKA, K. (1997) Trophic ecomorphology of Tanganyikan cichlids. Pp. 25-56 in: H. Kawanabe, M. Hori, & M. Nagoshi (Eds.). *Fish communities in Lake Tanganyika*. Kyoto Univ. Press, Kyoto, Japan.

YANAGISAWA, Y. (1985) Parental strategy of the cichlid fish *Perissodus microlepis*, with particular reference to intraspecific brood farming out. *Env. Biol. Eishes*, 12: 241-249.

YANAGISAWA, Y. (1986) Parental care in a monogamous mouthbrooding cichlid *Xenotilapia flavipinnis* in Lake Tanganyika. *Jap. J. Ichthyol.*, 33: 249-261.

YANAGISAWA, Y. & M. NISHIDA (1991) The social and mating system of the maternal mouthbrooder *Tropheus moorii* (Cichlidae) in Lake Tanganyika. *Japan. J. Ichthyol.*, 38: 271-282.

YANAGISAWA, Y. & M. NSHOMBO (1983) Reproduction and parental care of the scale-eating cichlid fish *Perissodus microlepis* in Lake Tanganyika. *Physiol. Ecol. Japan*, 20, pp 23-31.

YANAGISAWA, Y., H. OCHI & A. ROSSITER (1996) Intra-buccal feeding of young of in an undescribed Tanganyikan cichlid *Microdontochromis* sp. *Env. Biol. Fish.* 47: 191-201

YANAGISAWA, Y. & T. SATO (1990) Active browsing by mouthbrooding females of *Tropheus duboisi* and *Tropheus moorii* (Cichlidae) to feed the young and/or themselves. *Env. Biol. Fish.*, 27: 43- 50.

YUMA, M. & T. KONDO (1997) Interspecific relationships and habitat utilization among benthivorous cichlids. Pp. 87- 104 in: H. Kawanabe, M. Hori, & M. Nagoshi (Eds.). *Fish communities in Lake Tanganyika*. Kyoto Univ. Press, Kyoto, Japan.

Index